Karl Goser · Peter Glösekötter · Jan Dienstuhl

Nanoelectronics and Nanosystems

Springer
*Berlin
Heidelberg
New York
Hongkong
London
Mailand
Paris
Tokio*

Karl Goser · Peter Glösekötter · Jan Dienstuhl

# Nanoelectronics and Nanosystems

From Transistors to Molecular and Quantum Devices

With 254 Figures

 Springer

Professor em. Dr.-Ing. Karl Goser
Dipl.-Ing. Jan Dienstuhl
University of Dortmund
Faculty of Electrical Engineering & Information Technology
Integrated Systems Institute
44227 Dortmund
Germany

Dr.-Ing. Peter Glösekötter
Intel Corporation
Optical Components Division
Theodor-Heuss-Str. 7
38122 Braunschweig
Germany

Cataloging-in-Publication Data applied for
Bibliographic information published by Die Deutsche Bibliothek
Die Deutsche Bibliothek lists this publication in the Deutsche Nationalbibliografie;
detailed bibliographic data is available in the Internet at <http://dnb.dd.de>

ISBN 3-540-40443-0 Springer-Verlag Berlin Heidelberg New York

This work is subject to copyright. All rights are reserved, whether the whole or part of the material is concerned, specifically the rights of translation, reprinting, reuse of illustrations, recitation broadcasting, reproduction on microfilm or in other ways, and storage in data banks. Duplication o this publication or parts thereof is permitted only under the provisions of the German Copyright Law of September 9, 1965, in its current version, and permission for use must always be obtained from Springer-Verlag. Violations are liable for prosecution under German Copyright Law.

Springer-Verlag is a part of Springer Science+Business Media

springeronline.com

© Springer-Verlag Berlin Heidelberg 2004
Printed in Germany

The use of general descriptive names, registered names, trademarks, etc. in this publication does no imply, even in the absence of a specific statement, that such names are exempt from the relevan protective laws and regulations and therefore free for general use.

Typesetting: Digital data supplied by authors
Cover-Design: Design & Production, Heidelberg
Printed on acid-free paper    62/3020 Rw 5 4 3 2 1 0

# Preface

In recent years nanoelectronics has been rapidly gaining in importance and is already on the way to continuing the outstanding success of microelectronics. While most literature dealing with nanoelectronics is concerned with technology and devices, it has been nearly impossible to find anything about the circuit and system level. The challenges of nanoelectronics, however, are evident not only in the manufacture of tiny structures and sophisticated nanodevices, but also in the development of innovative system architectures that will have to orchestrate billions of devices inside future gadgets. This book's objective is to bridge that gap.

The topic of this book has actually been offered as a student's lecture at the University of Dortmund. Since the lecture has been held at the faculty of engineering, the main focus lies on electronics and on the basic principles of the essential physical phenomena.

This book represents an introduction to nanoelectronics, as well as giving an overview of several different technologies and covering all aspects from technology to system design. On the system level, various architectures are presented and important system features - namely scalability, processing power and reliability - are discussed. A variety of different technologies are presented which include molecular, quantum electronic, resonant tunnelling, single-electron and superconducting devices and even devices for DNA and quantum computing. Additionally, the book encompasses a comparison between nanoelectronics and the present state of silicon technologies, a discussion of the nanoelectronic limits, and a vision of future nanosystems in terms of information technologies.

This book is intended for those people who have not lost sight of the system as a whole. It not only covers nano-technology and its devices, but also considers the system and circuit level perspective, indicating the applications and conceivable products. The overview is written for electronic engineers, computer scientists, marketing people and physicists, and offers an initial orientation for students, beginners and outsiders alike.

Several EU projects and various EU initiatives form the scientific basis for the current research activities in the fields of nanoelectronics. We would therefore like to thank the co-ordinator in Brussels, Romano Compano, for his helpful undertakings. In addition we would like to thank our partners at Infineon Research Laboratory in Munich for our many fruitful discussions. Thanks are also due to the Deutsche Forschungsgemeinschaft, especially for the support of the Collaborative Research Center 531, which deals with the design and management of complex technical processes and systems by means of Computational Intelligence methods.

We also thank the assistants and colleagues who coached the lecture: C. Burwick, A. Kanstein, G. Wirth, M. Rossmann and C. Pacha. Thanks are also due to the assistants who drew the diagrams: T. Kliem, G. Sapsford, K. Möschke, B. Rückstein. Last, but not least, thanks go to Karin Goser who corrected the German and English versions.

Dortmund,
July 2003

*Karl F. Goser*
*Peter Glösekötter*
*Jan Dienstuhl*

# Contents

| | | |
|---|---|---|
| **1** | **On the Way to Nanoelectronics** | 1 |
| | 1.1 The Development of Microelectronics | 2 |
| | 1.2 The Region of Nanostructures | 4 |
| | 1.3 The Complexity Problem | 7 |
| | 1.4 The Challenge initiated by Nanoelectronics | 9 |
| | 1.5 Summary | 11 |
| **2** | **Potentials of Silicon Technology** | 13 |
| | 2.1 Semiconductor as Base Material | 13 |
| |     2.1.1 Band Diagram of a Semiconductor | 13 |
| |     2.1.2 Band Diagrams of Inhomogeneous Semiconductor Structures | 15 |
| | 2.2 Technologies | 16 |
| |     2.2.1 Different Types of Transistor Integration | 17 |
| |     2.2.2 Technological Processes for Microminiaturization | 19 |
| | 2.3 Methods and Limits of Microminiaturization in Silicon | 23 |
| |     2.3.1 Scaling | 23 |
| |     2.3.2 Milestones of Silicon Technology | 24 |
| |     2.3.3 Estimation of Technology Limits | 26 |
| | 2.4 Microelectronic and Mechanical Systems (MEMS) | 31 |
| |     2.4.1 Technology of Micromechanics | 31 |
| |     2.4.2 Micromechanics for Nanoelectronics | 32 |
| | 2.5 Integrated Optoelectronics | 34 |
| | 2.6 Conclusion | 37 |
| **3** | **Basics of Nanoelectronics** | 39 |
| | 3.1 Some Physical Fundamentals | 39 |
| |     3.1.1 Electromagnetic Fields and Photons | 39 |
| |     3.1.2 Quantization of Action, Charge, and Flux | 41 |
| |     3.1.3 Electrons Behaving as Waves (Schrödinger Equation) | 42 |
| |     3.1.4 Electrons in Potential Wells | 45 |

VIII   Contents

|  |  | 3.1.5 Photons interacting with Electrons in Solids . . . . . . . . . . | 46 |
|---|---|---|---|

        3.1.5  Photons interacting with Electrons in Solids . . . . . . . . . . 46
        3.1.6  Diffusion Processes . . . . . . . . . . . . . . . . . . . . . . . . . . . . . . . 48
  3.2  Basics of Information Theory . . . . . . . . . . . . . . . . . . . . . . . . . . . . . 52
        3.2.1  Data and Bits . . . . . . . . . . . . . . . . . . . . . . . . . . . . . . . . . . . 52
        3.2.2  Data Processing . . . . . . . . . . . . . . . . . . . . . . . . . . . . . . . . . 56
  3.3  Summary . . . . . . . . . . . . . . . . . . . . . . . . . . . . . . . . . . . . . . . . . . . . . 59

**4 Biology-Inspired Concepts** . . . . . . . . . . . . . . . . . . . . . . . . . . . . . . . 61
  4.1  Biological Networks . . . . . . . . . . . . . . . . . . . . . . . . . . . . . . . . . . . . 61
        4.1.1  Biological Neurons . . . . . . . . . . . . . . . . . . . . . . . . . . . . . . . 61
        4.1.2  The Function of a Neuronal Cell . . . . . . . . . . . . . . . . . . . . 64
  4.2  Biology-Inspired Concepts . . . . . . . . . . . . . . . . . . . . . . . . . . . . . . . 67
        4.2.1  Biological Neuronal Cells on Silicon . . . . . . . . . . . . . . . . . . 68
        4.2.2  Modelling of Neuronal Cells by VLSI Circuits . . . . . . . . . 69
        4.2.3  Neuronal Networks with local Adaptation and
               Distributed Data Processing . . . . . . . . . . . . . . . . . . . . . . . . 72
  4.3  Summary . . . . . . . . . . . . . . . . . . . . . . . . . . . . . . . . . . . . . . . . . . . . . 75

**5 Biochemical and Quantum-mechanical Computers** . . . . . . . . . 77
  5.1  DNA Computer . . . . . . . . . . . . . . . . . . . . . . . . . . . . . . . . . . . . . . . 78
        5.1.1  Information Processing with Chemical Reactions . . . . . . 78
        5.1.2  Nanomachines . . . . . . . . . . . . . . . . . . . . . . . . . . . . . . . . . . . 79
        5.1.3  Parallel Processing . . . . . . . . . . . . . . . . . . . . . . . . . . . . . . . 82
  5.2  Quantum Computer . . . . . . . . . . . . . . . . . . . . . . . . . . . . . . . . . . . . 83
        5.2.1  Bit and Qubit . . . . . . . . . . . . . . . . . . . . . . . . . . . . . . . . . . . 83
        5.2.2  Coherence and Entanglement . . . . . . . . . . . . . . . . . . . . . . 85
        5.2.3  Quantum Parallelism . . . . . . . . . . . . . . . . . . . . . . . . . . . . . 86
  5.3  Summary . . . . . . . . . . . . . . . . . . . . . . . . . . . . . . . . . . . . . . . . . . . . . 88

**6 Parallel Architectures for Nanosystems** . . . . . . . . . . . . . . . . . . . 89
  6.1  Architectural Principles . . . . . . . . . . . . . . . . . . . . . . . . . . . . . . . . . 89
        6.1.1  Mono- and Multiprocessor Systems . . . . . . . . . . . . . . . . . 89
        6.1.2  Some Considerations to Parallel Data Processing . . . . . . 91
        6.1.3  Influence of Delay Time . . . . . . . . . . . . . . . . . . . . . . . . . . 92
        6.1.4  Power Dissipation and Parallelism . . . . . . . . . . . . . . . . . . 95
  6.2  Architectures for Parallel Processing in Nanosystems . . . . . . . . . 97
        6.2.1  Classic Systolic Arrays . . . . . . . . . . . . . . . . . . . . . . . . . . . . 97
        6.2.2  Processors with Large Memories . . . . . . . . . . . . . . . . . . . . 98
        6.2.3  Processor Array with SIMD and PIP Architecture . . . . . 100
        6.2.4  Reconfigurable Computer . . . . . . . . . . . . . . . . . . . . . . . . . 101
        6.2.5  The Teramac Concept as a Prototype . . . . . . . . . . . . . . . 101
  6.3  Summary . . . . . . . . . . . . . . . . . . . . . . . . . . . . . . . . . . . . . . . . . . . . 104

## Contents IX

**7  Softcomputing and Nanoelectronics** .......................... 107
   7.1  Methods of Softcomputing ................................ 108
       7.1.1  Fuzzy Systems ..................................... 108
       7.1.2  Evolutionary Algorithms ........................... 112
       7.1.3  Connectionistic Systems ........................... 113
       7.1.4  Computational Intelligence Systems ................ 115
   7.2  Characteristics of Neural Networks in Nanoelectronics........ 117
       7.2.1  Local Processing .................................. 117
       7.2.2  Distributed and Fault-Tolerant Storage ............. 118
       7.2.3  Self-Organization ................................. 120
   7.3  Summary ............................................... 122

**8  Complex Integrated Systems and their Properties** ......... 123
   8.1  Nanosystems as Information-Processing Machines .......... 123
       8.1.1  Nanosystems as Functional Blocks .................. 123
       8.1.2  Information Processing as Information Modification ... 124
   8.2  System Design and its Interfaces ......................... 126
   8.3  Evolutionary Hardware ................................. 129
   8.4  Requirements of Nanosystems ........................... 130
   8.5  Summary ............................................... 132

**9  Integrated Switches and Basic Circuits** .................... 133
   9.1  Switches and Wiring .................................... 134
       9.1.1  Ideal and Real Switches ........................... 134
       9.1.2  Ideal and Real Wiring ............................. 137
   9.2  Classic Integrated Switches and their Basic Circuits.......... 141
       9.2.1  Example of a Classic Switch: The Transistor ......... 141
       9.2.2  Conventional Basic Circuits ........................ 142
       9.2.3  Threshold Gates .................................. 145
       9.2.4  Fredkin Gate ..................................... 147
   9.3  Summary ............................................... 149

**10  Quantum Electronics** ....................................... 151
   10.1  Quantum Electronic Devices (QED) ..................... 151
       10.1.1  Upcoming Electronic Devices ..................... 151
       10.1.2  Electrons in Mesoscopic Structures ................ 153
   10.2  Examples of Quantum Electronic Devices ................ 156
       10.2.1  Short-Channel MOS Transistor ................... 156
       10.2.2  Split-Gate Transistor ............................ 157
       10.2.3  Electron-Wave Transistor ........................ 158
       10.2.4  Electron-Spin Transistor ......................... 159
       10.2.5  Quantum Cellular Automata (QCA) ............... 160
       10.2.6  Quantum-Dot Array ............................. 165
   10.3  Summary .............................................. 166

## 11 Bioelectronics and Molecular Electronics ................... 169
### 11.1 Bioelectronics .............................................. 170
#### 11.1.1 Molecular Processor ................................. 171
#### 11.1.2 DNA Analyzer as Biochip ........................... 172
### 11.2 Molecular Electronics ..................................... 174
#### 11.2.1 Overview ............................................. 174
#### 11.2.2 Switches based on Fullerenes and Nanotubes .......... 175
#### 11.2.3 Polymer Electronic ................................... 178
#### 11.2.4 Self-Assembling Circuits ............................. 180
#### 11.2.5 Optical Molecular Memories ......................... 182
### 11.3 Summary .................................................. 185

## 12 Nanoelectronics with Tunneling Devices ................... 187
### 12.1 Tunneling Element (TE) ................................... 187
#### 12.1.1 Tunnel Effect and Tunneling Elements ............... 188
#### 12.1.2 Tunneling Diode (TD) ............................... 190
#### 12.1.3 Resonant Tunneling Diode (RTD) .................... 192
#### 12.1.4 Three-Terminal Resonant Tunneling Devices .......... 196
### 12.2 Technology of RTD ........................................ 196
### 12.3 Digital Circuit Design Based on RTDs ...................... 198
#### 12.3.1 Memory Applications ................................. 198
#### 12.3.2 Basic Logic Circuits .................................. 198
#### 12.3.3 Dynamic Logic Gates ................................. 199
### 12.4 Digital Circuit Design Based on the RTBT ................. 204
#### 12.4.1 RTBT MOBILE ..................................... 204
#### 12.4.2 RTBT Threshold Gate ............................... 205
#### 12.4.3 RTBT Multiplexer ................................... 206
### 12.5 Summary .................................................. 208

## 13 Single-Electron Transistor (SET) .......................... 209
### 13.1 Principle of the Single-Electron Transistor .................. 209
#### 13.1.1 The Coulomb Blockade .............................. 209
#### 13.1.2 Performance of the Single-Electron Transistor ......... 211
#### 13.1.3 Technology .......................................... 214
### 13.2 SET Circuit Design ........................................ 216
#### 13.2.1 Wiring and Drivers .................................. 216
#### 13.2.2 Logic and Memory Circuits .......................... 217
#### 13.2.3 SET Adder as an Example of a Distributed Circuit .... 220
### 13.3 Comparison Between FET and SET Circuit Designs ......... 220
### 13.4 Summary .................................................. 223

## 14 Nanoelectronics with Superconducting Devices ............ 225
- 14.1 Basics ............................................. 225
  - 14.1.1 Macroscopic Characteristics ...................... 225
  - 14.1.2 The Macroscopic Model ........................... 227
- 14.2 Superconducting Switching Devices ..................... 228
  - 14.2.1 Cryotron ........................................ 228
  - 14.2.2 The Josephson Tunneling Device .................. 229
- 14.3 Elementary Circuits .................................. 231
  - 14.3.1 Memory Cell ..................................... 231
  - 14.3.2 Associative or Content-Addressable Memory ........ 232
  - 14.3.3 SQUID - Superconducting Quantum Interferometer Device ........................................... 233
- 14.4 Flux Quantum Device ................................. 233
  - 14.4.1 LC-Gate ......................................... 234
  - 14.4.2 Magnetic Flux Quantum - Quantum Cellular Automata 234
  - 14.4.3 Quantum Computer with Single-Flux Devices ........ 234
  - 14.4.4 Single Flux Quantum Device - SFQD ............... 236
  - 14.4.5 Rapid Single Flux Quantum Device - RSFQD ........ 237
- 14.5 Application of Superconducting Devices ................ 238
  - 14.5.1 Integrated Electronics ........................... 238
  - 14.5.2 FET Electronics - A Comparison .................. 239
  - 14.5.3 The Electrical Standards ......................... 241
- 14.6 Summary ............................................ 242

## 15 The Limits of Integrated Electronics ..................... 245
- 15.1 A Survey about the Limits ............................ 245
- 15.2 The Replacement of Technologies ...................... 246
- 15.3 Energy Supply and Heat Dissipation ................... 248
- 15.4 Parameter Spread as Limiting Effect ................... 252
- 15.5 The Limits due to Thermal Particle Motion ............. 257
  - 15.5.1 The Debye Length ................................ 257
  - 15.5.2 Thermal Noise ................................... 258
- 15.6 Reliability as Limiting Factor ........................ 259
- 15.7 Physical Limits ..................................... 263
  - 15.7.1 Thermodynamic Limits ............................ 264
  - 15.7.2 Relativistic Limits .............................. 264
  - 15.7.3 Quantum-Mechanical Limits ....................... 265
  - 15.7.4 Equal Failure Rates by Tunneling and Thermal Noise .. 265
- 15.8 Summary ............................................ 266

## 16 Final Objectives of Integrated Electronic Systems ......... 267
- 16.1 Removal of Uncertainties by Nanomachines ............. 267
- 16.2 Uncertainties in Nanosystems ......................... 269
- 16.3 Uncertainties in the Development of Nanoelectronics ... 270
- 16.4 Summary ............................................ 271

**References** . . . . . . . . . . . . . . . . . . . . . . . . . . . . . . . . . . . . . . . . . . . . . . . . . . . . . . 273

**Index** . . . . . . . . . . . . . . . . . . . . . . . . . . . . . . . . . . . . . . . . . . . . . . . . . . . . . . . . . . . 277

# 1
## On the Way to Nanoelectronics

The experience of the past shows that throughout constant technology improvement electronics has become more reliable, faster, more powerful, and less expensive by reducing the dimensions of integrated circuits. These advantages are the driver for the development of modern microelectronics. The long-term goal of this development will lead to nanoelectronics. The first microelectronic components and systems were quite expensive and therefore only an adequate solution for space travel. Nowadays integrated circuits as key components, are utilized in a broad range of applications. The semiconductor silicon is the most important material in the production of microelectronic circuits. Actually the limit of silicon technology is set by the manufacturing processes and not by silicon itself or the laws of physics.

In the beginning of microelectronics in 1960 the way of its development was not clear. It was absolutely not obvious that the circuits must be integrated into silicon, however, it was quite obvious to use solid state switches. The invention of the bipolar transistor (Bardeen, Brittain, and Shockley in 1948) was only possible through extensive studies in the field of solid-state materials some decades before. This pioneering work was awarded the Nobel prize. It was crucial that both the Bell Telephone Company and the inventors shared their knowledge with the public and the further development was not restricted by patents. The invention of the integrated circuit by Kilby and Noice in 1959 was also spectacular, however, it was quite obvious to integrate both transistors and resistors on a single chip. For this work Kilby was awarded the 2000 Nobel prize. As we know today his idea was the right approach for the development of microelectronic circuits. Other ideas were less spectacular, however, they also influenced the spread of microelectronics. Some examples of these ideas are the planar technology, the microprocessor concept (Hoff 1971 with the Intel 4004), the scaling of MOS circuits (Dennard 1973) and the technology-invariant interface of the design rules (Mead and Conway 1980) [1].

The significance of microelectronics as a basic technology was hesitatingly accepted by the market. Nowadays we recognize microelectronics as an important key technology for present and future information systems. In particular

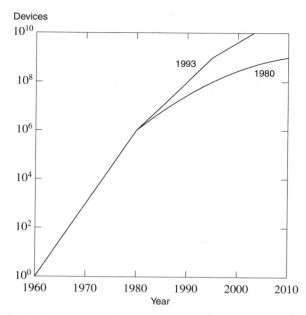

**Fig. 1.1.** Integration level of memory chips as the highest integrated circuits. In the past the integration level doubled every 18 months. The prediction by J. Meindl in 1980 was superseded by the curve of 1993, the development moves faster than expected

the growing demands of information technology for more powerful microelectronic circuits will enforce the transition to nanoelectronics in the future.

## 1.1 The Development of Microelectronics

The progress in microelectronics is indicated by the integration level over the last decades of development. Figure 1.1 shows a curve that J. Meindl forecasted in 1980 [2]. Gordon Moore, one of the founders of Intel, proposed this representation. The Moore Plot shows that the integration level is doubling every 18 months. Figure 1.1 shows two tendencies, both are too reserved about the real development, so the curves have to be corrected to higher values [3]. Today we have the vision of memory chips made of silicon with a capacity of 160 Gbit (4 Gbit can store a whole movie).

Such trend curves of microelectronics show the essential characteristics of the devices, systems, or other parameters and can be extrapolated for future predictions. These "laws" are not laws of physics, they are mainly based on the laws of business management. They are valid because many people are working under the same economical conditions and technical prerequisites, and under the same social stability. Under these conditions progress is continuousl. In

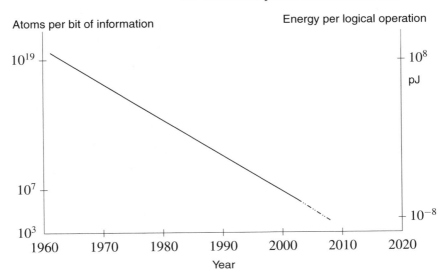

**Fig. 1.2.** The development of integrated circuits shows changes in the orders of magnitude. The number of atoms and the switching energy necessary for operating one bit are reduced by several powers of ten

addition, the above-mentioned is only valid under the precondition that the scientific laws allow such a progress. At the moment the limits set by the laws of nature are relatively still far away, as will be shown in Chap. 15.

Since the invention of integrated circuits we can see changes of their characteristics in the orders of magnitude (Fig. 1.2). If we regard the number of atoms that are necessary for storing or operating 1 bit then we can see the breath-taking development: In 1960 one transistor consisted of $10^{20}$ atoms in a volume of $0.1\,\mathrm{cm}^3$, in the year 2000 these numbers were reduced to $10^7$ atoms in $0.01\,\mu\mathrm{m}^3$. In the same way the energy for storing or operating 1 bit decreased, since the energy for charging and discharging capacities was lowered by the facts that one reduces the area of the capacitors from $1\,\mathrm{cm}^2$ to $0.01\,\mu\mathrm{m}^2$ and the voltages from 10 V to 1 V. The change of 12 orders of magnitude was the prerequisite that microelectronics could have such high integration levels without running into the problems of power dissipation and thermal heating.

The increase of the integration level was possible first of all by reducing the feature size of the devices and in the second place by enlarging the chip area and by functional integration (Fig. 1.3). It is assumed that the silicon technology has its limits for further miniaturization to approximately 10 nm in 2010. To overcome this restriction, functional integration and three-dimensional integration are possible solutions. An alternative approach is nanoelectronics.

Nanoelectronics will offer integration levels that will be about two to four orders higher than those of microelectronics by 2010. Therefore we can expect

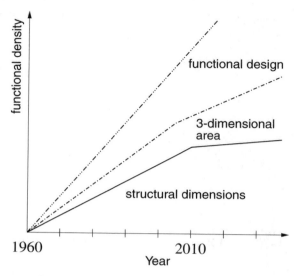

**Fig. 1.3.** Increase of the integration level over the years via three ways: smaller structures, larger chip areas, and functional design

memories with capacities up to one Terabit. Nanoelectronics is characterized by very small structures and by very high complexities. The following chapters will discuss both features.

## 1.2 The Region of Nanostructures

Figure 1.4 shows different physical levels if the feature sizes of the devices are scaled down. The characteristic times in semiconductors are plotted against the structural dimensions of the devices. These times are correlated to different physical effects. The shaded area covers present-day microelectronics. Outside this region the domain of nanoelectronics and molecular electronics is reached. The characteristic times and dimensions are less than 1 ps and less than 10 nm respectively.

For example, the depletion-layer width limits the reduction in size of the pn-junction in a diode [4]. For a $p^+$n-junction the depletion width can be approximated by

$$l_n = \sqrt{\frac{2\varepsilon\,(V_D - V)}{q\,N_D}}. \tag{1.1}$$

The width $l_n$ of the depletion layer increases with the voltage $V$ and decreases with the doping density $N_D$ of the region. The permittivity is denoted by $\varepsilon$ and $q$ is the elementary charge. Another important dimension is the Debye length

1.2 The Region of Nanostructures    5

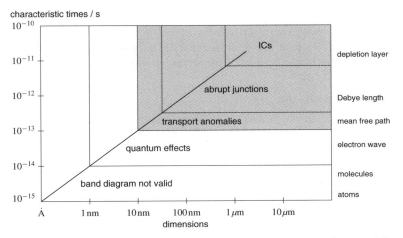

**Fig. 1.4.** Characteristic times and structures for semiconductor devices. The grey area concerns the present integrated circuits

$$L_D = \sqrt{\frac{\epsilon V_T}{q\, n_i}}. \tag{1.2}$$

The definition of the Debye length is similar to that of the depletion-layer width. In (1.1) the voltage is replaced by the thermal voltage $V_T = \frac{kT}{q}$ and the doping density $N_D$ by the intrinsic doping $n_i$ of a semiconductor. For a doped semiconductor $n_i$ is substituted by the density of the majority carriers. The Debye length describes the spatial extension of a perturbation inside a semiconductor.

Particles moving due to thermal energy have a mean free path $\bar{l}$ between two collisions

$$\bar{l} = \overline{v_{th} \tau_c}. \tag{1.3}$$

Equation (1.3) depends on the thermal velocity $\overline{v_{th}}$ and on the mean free time $\overline{\tau_c}$ of the particles, which determines the mobility $\mu$ of the charge carriers.

Quantum effects become relevant in devices if the wavelength $\lambda$ of the electrons is in the range of the feature size of the devices.

$$\lambda = \frac{h}{m_e v}. \tag{1.4}$$

Equation (1.4) $h$ is Planck's constant. The wavelength $\lambda$ is inversely proportional to the mass of the particle $m_e$ and the velocity $v$.

Most of the nanoelectronic devices function in this range so the wave behavior of the electrons has to be considered. If one continues to decrease the structural dimension the domain of atoms and molecules is reached. If a nanoelectronic device has only one structural dimension in this range we call it a *mesoscopic* device (meso means in between). Due to the disturbing influence of the thermal energy many quantum devices are operated at low temperatures. However, these quantum effects increase with decreasing feature size of the devices. Therefore the devices must be very small if operated at room temperature. The latter is also an important point for nanoelectronics.

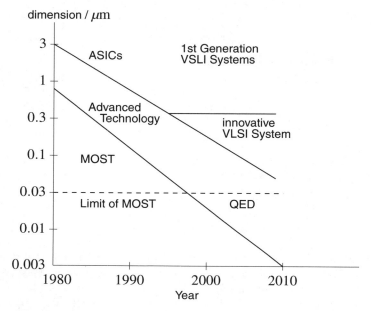

**Fig. 1.5.** The development of the feature size of chips, the limit of MOS transistors was predicted as 1995, today we assume the limit at 10 nm

Figure 1.5 shows the reduction of the structure sizes with regard to the year of development for application specific integrated circuits (ASICs) and for state-of-the-art technology research samples. The development of the silicon technology is summarized by the Semiconductor Industries Association (SIA) roadmap, which is widely accepted by experts [5]. Today the roadmap predicts a minimum channel length of MOS transistors in the order of 10 nm, some years ago it was 30 nm. Below this size quantum electronic devices (QED) and mesoscopic devices will start to replace MOS transistors.

The development of higher integration levels forces two conclusions: First, new physical effects, for example quantum effects due to small dimensions, replace the classical transistors. Second, the huge number of devices on a single chip demands new system concepts due to the reduced efficiency of the classic architectures. The architectures must consider fault tolerance, self-organization, and they must satisfy a large variety of applications. If silicon technology will eventually reach its limits other technologies must be applied, unless there is no demand from the market to introduce other technologies with improved characteristics. However, this is unlikely because the demands of the information technology will be tremendous.

Another limit may arise from economics when silicon technology is too cost-intensive. It is assumed that in 10 years the sale of chips will be equal to the sale of electronics. In this case the chips will be too expensive and not economic. If a new technology is more economic than the last one then the market will accept the new technology. The experiences gained from the past show that small structures mean low costs, so nanoelectronics may be the right way from the economic point of view, too.

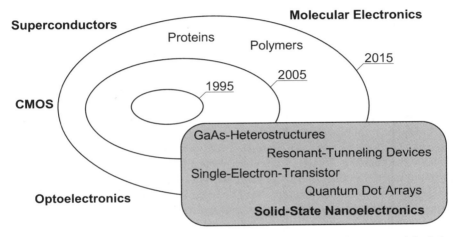

**Fig. 1.6.** Diversity of technologies of today and the future. Their potential of development is indicated

Figure 1.6 gives a rough overview of the most important technologies of the future. A new technology must be better in performance and less in costs if it is to be accepted by the market. Today the choice of a specific technology is still open. The following chapters will describe a small selection of possible technologies for nanoelectronic circuits, a final conclusion is given in Chap. 16.

## 1.3 The Complexity Problem

In addition to the small structures, in the order of some nanometers, another point will be of great interest: the high complexity of integrated nanoelectronic systems. Modern microelectronic systems contain up to 100 million devices on a single chip. Nanoelectronics will push this number up to 1 billion devices or even more. The main problem is not only the large number of devices, but also the development time and the time for testing such systems. Another important point of view is the choice of an architecture for an efficient interaction between the subsystems.

The degree of complexity can be analyzed by combinatorial mathematics. Microelectronics distinguishs between an implicit and explicit complexity. Memory chips, as an example for regular circuits, are characterized by many data bits: they show a high explicit complexity. Examples for an implicit complexity are microprocessors with their instruction sets, analog circuits with their complex behavior, or the wiring of a large circuit.

The analysis of memory chips gives a starting point for this problem. Testing of a memory chip with $n$ memory cells, in which each cell can store 1 bit requires $2n$ test cycles for reading and writing 0 s and 1 s. A 16 Mbit memory chip therefore requires 32 million test cycles. Such simple test patterns are not sufficient for testing all possible errors of a memory chip because the cells can not be regarded independently of one another. If we intend to examine all possible dependencies the test must cover all possible bit patterns that can be stored on the chip. A chip with $n$ cells has $2^n$ different bit patterns, a chip with 16 777 216 cells requires about $2^{16\ 777\ 216} \approx 10^{5\ 000\ 000}$ test cycles. Assuming 1 ns for each cycle the testing of the whole chips lasts longer than the age of the universe. We can see that time is of major concern even from the economic point of view. This problem will rise exponentially with increasing complexity of the chips. In practice test engineering restricts itself to the sensitive cases in regard to the interference between cells. The price to be paid for this pragmatic solution is memory chips that can have soft failures. This problem can be diminished by special software for error correction.

A further interesting example for the complexity in microelectronics is the wiring of an integrated circuit. The task is to minimize the length of the signal or bus lines, a very complex problem. If $n$ denotes the number of points or devices that should be connected there are $m = n!$ possibilities for connecting them. Problems of this nature are said to be NP complete problems. This problem is more complex than testing of a chip and is well known as the *traveling salesman problem* (TSP). The verification of a possible solution is as difficult as the problem itself. Parallel computing with present-day microelectronics is not capable of finding a solution for this problem for large values of $n$. These problems will force the development of nanoelectronics.

The last examples show that if $n$ becomes large $m$ becomes very large, for simplicity the logarithm is introduced. An example is (1.5) for the definition of information content :

$$H_0 = \operatorname{ld} m. \qquad (1.5)$$

The information content $H_0$ defines the level of complexity in bits if we apply the logarithm to the base of two. The symbol of this logarithm is ld (logarithmus dualis). Similar to thermodynamics that describes the number of microstates for the distribution of particles in a volume the information content is denoted as *negentropy* (negative entropy). In the following we give some examples for the negentropy of a different system: A card game has $H_0 = 3.2 \times 10^2$ bit, a microprocessor composed of 10 000 transistors $H_0 = 5.4 \times 10^6$ bit, a telephone network for 10 millions of participants

$H_0 = 2.3 \times 10^8$ bit. A television picture consists of $10^8$ bit. Larger values for $H_0$ offer artificial neural networks. A net with $10^4 \times 10^4$ cells has $H_0 = 2.9 \times 10^9$ bit, whereas a brain with $10^{14}$ neurons has $H_0 = 4.6 \times 10^{15}$ bit if this simple model is applied.

A comparison with entropy in thermodynamics is interesting. The entropy is defined by
$$S = k \cdot \ln m. \tag{1.6}$$

In this relation $k$ is Boltzmann's constant and $m$ the number of microstates. The values that occur in thermodynamics are much larger than those of information technology. If we heat $1\,\text{cm}^3$ of silicon from $20°C$ to $21°C$ and apply (1.6) we get $S = 3.76\,\text{J}\,\text{K}^{-1}$ or applying (1.5) we get $H_0 = 3.9 \times 10^{23}$ bit. Such large values are important to everyday life because mean values, e.g. the temperature varies between the mean value very precisely. On the other hand, information processing is capable of optimizing the thermal energy in our houses because the energy of a microprocessor is negligible compared to the energy of the controlled system.

## 1.4 The Challenge initiated by Nanoelectronics

The quantity of information is continuously increasing. The exponential growth of the data and knowledge is based on new scientific discoveries and particularly on the public sector. Many of our technical systems become more and more complex, e.g. systems for environmental protection will be widely used in future. These systems demand components with high-performance information processing. In general, a system is defined as an assemblage or a combination of things or parts forming a complex or unitary whole. In the following we will concentrate on the architecture, the implementation, and the realization of integrated systems for all forms of information processing.

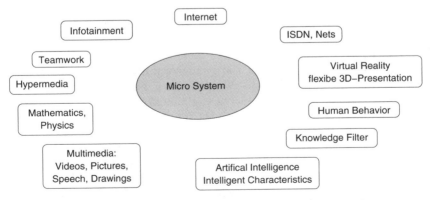

**Fig. 1.7.** Functional diagram of brilliant systems for PCs

An interesting example for an innovative product is a brilliant system that has been launched on the market in a very simple version. Such systems offer the following features (Fig. 1.7): Communication with the Internet, infotainment can help to understand difficult contents, e.g. mathematical problems. For daily work, hypermedia, multimedia, and team-working are important characteristics. Additional topics are three-dimensional presentations, virtual realities, and knowledge filters. The concept of a silent servant working in the background is capable of collecting and storing relevant data, and protecting the system against misuse. Intelligent firewall systems prevent the systems from viruses.

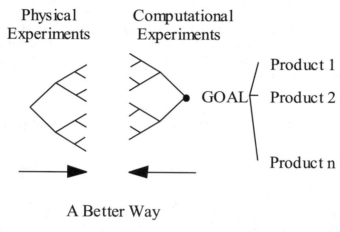

**Fig. 1.8.** The goals of nanoelectronics should be tackled not only from the side of technology but also from system engineering. The best strategies meet in the middle, and the goal is product-oriented development

The continuous increase in performance for data-processing tasks is the main driver for microelectronic systems. In the past the performance of a system expressed in terms of the measure *MIPS* (million instructions per second) has increased every 10 years by a factor of ten. This increase is based on a continuous improvement of the technology in microelectronics. A different approach for increasing the performance is based on the choice of the architecture. Parallel processing increases the performance of data processing without changing the technology. This concept does not fulfil all claims. New concepts such as, for example, artificial neural networks with the important characteristic of selforganization of its structures seem to be very promising.

The experiences gained in microelectronics show that the development of new technologies is not only a matter of material science but also of product engineering. Therefore the development of new systems is a joint task between technology development and system engineering, as Fig. 1.8 indicates. The development includes the concepts of top-down and bottom-up.

## 1.5 Summary

Today the driving force for microelectronics is information technology. The challenges of the future will demand sophisticated microelectronics, probably nanoelectronics. The development of reasonable concepts is a challenge for engineering and computer science, and an interdisciplinary task, probably for a new field that may be called information technology.

# 2
# Potentials of Silicon Technology

Nanoelectronics has to be discussed in the background of today's microelectronic capabilities. Silicon is the predominant material of present microelectronic technologies, since it has very suitable integration properties. However, other materials such as GaAs have even better properties. Among other things, the already gathered production knowledge of silicon-based technologies represents a huge capital expenditure that still has to be found for other technologies. Thus, alternative technologies are interesting and provide important information for the further development of microelectronics, but at present they do not show any economic importance in comparison with silicon technology.

## 2.1 Semiconductor as Base Material

### 2.1.1 Band Diagram of a Semiconductor

The most important properties of semiconductors derive from the quantum-mechanical models (see Chap. 3). The physical shape of a body, but also its inside structure has an impact on the electron wave. The band diagram derives from the nanometric crystal structure of the semiconductor.

Within the crystal electrons are ruled by a periodic potential (Fig. 2.1a). It can be considered as periodic lined-up potential wells that are separated from each other by potential barriers. In this case, the $\Psi$ function can be described by Schrödinger's equation that shows allowed energy bands as well as bandgaps in the E(k)-diagram. Figure 2.1b shows the E(x)-diagram, which is more appropriate for practical purposes, since x refers to the x-dimension of the semiconductor. Because of the quantum wells, electrons can only take discrete energy levels. The potential wells are coupled with each other via the tunnel effect that results in a widening of the discrete energy levels. They form energy bands (Chap. 3).

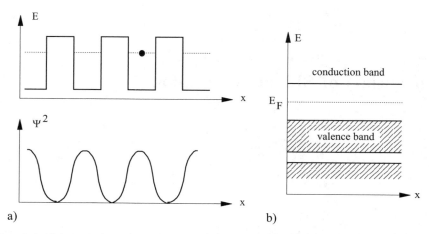

**Fig. 2.1.** (a) Periodic potential of an undoped crystal (nanometric domain) and its wave function, (b) The band diagram explains the behavior of electrons in the semiconductor

The E(x)-diagram comprises allowed energy bands that are separated by bandgaps. In terms of semiconductors, the uppermost allowed energy band is called the conduction band whereas the second uppermost allowed energy band is called the valence band. The conduction band comprises electrons as mobile charges that originate from the valence band; these defect electrons (holes) are modeled as positive charges of the valence band. The electron and hole density can be influenced via doping.

What is new about this is that the semiconductor is based on two types of charge, namely negative and positive (only virtual) charge. Both types coexist and contribute simultaneously to the conduction process. This behavior is of fundamental relevance in terms of inhomogeneous structures like the pn-junction.

An electric field E puts the force $F = qE$ on holes because of their elementary positive charge q. Due to thermal agitation, particles move slowly with the electric field. The mean effective velocity $V_{Fp}$ of the particles is proportional to the electric field E:

$$v_{Fp} = \mu_p E. \tag{2.1}$$

$\mu_p$ describes the mobility of the holes. This velocity is responsible within the semiconductor for both the charge density p and the current density $S_{Fp}$:

$$S_{Fp} = pq\mu_p E. \tag{2.2}$$

Since the electric field E is proportional to the applied voltage, (2.2) is equivalent to Ohm's law: The current density rises with the voltage in a linear manner. According to a relative simple model, the mobility $\mu_p$ depends on the

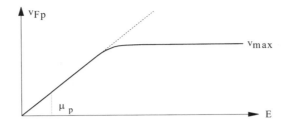

**Fig. 2.2.** Drift velocity as a function of the electric field. High electric fields cause a velocity saturation

effective mass $m^*$ and on the time $\tau_C$ that describes the mean time between two collisions with other particles or atoms of the crystal lattice:

$$\mu_p = \frac{v_{Fp}}{E} = \frac{q\tau_C}{m^*}. \qquad (2.3)$$

The mean free path is equivalent to $\tau_C v_{th}$. In this context $v_{th}$ describes the mean thermal velocity of the particles. $m^*$ as well as $\tau_C$ have to be extracted from the band diagram of the corresponding crystal.

The simple linear equation (2.2) only holds for relatively low electric fields. The interaction with the crystal lattice via collisions causes, for higher electric fields, a saturation of the drift velocity (Fig. 2.2). This saturation effect results in a substantial limitation for electronic systems in terms of their switching speed.

### 2.1.2 Band Diagrams of Inhomogeneous Semiconductor Structures

Figure 2.3 illustrates different inhomogeneous semiconductor structures and their corresponding energy-band model [6].

The starting point is the undoped semiconductor (a). Its Fermi level is approximately located in the middle of the bandgap. Doping shifts the Fermi level towards the band edges of the bandgap, which results in an n-type and a p-type semiconductor (b). Since the pn-junction (c) is an essential inhomogeneous structure, it appears in almost all semiconductor devices. The pn-junction degenerates to a tunneling diode if the doping levels are very high. The MOS structure (d) is based on an electric field that is applied vertically to the semiconductor surface. The potential well at the semiconductor surface can store charged particles. The bent band edges in the band diagram (d) reveal this potential well. The MOS transistor exploits this effect. The Schottky contact (e) is based on a metal-semiconductor junction and is utilized as contact and diode. Besides the potential shift the Schottky contact is also based on band bending. The metal-semiconductor-field-effect transistor (MESFET) (f) makes use of the Schottky contact. Further modifications of the band structure result from very thin and distinct doped semiconductor

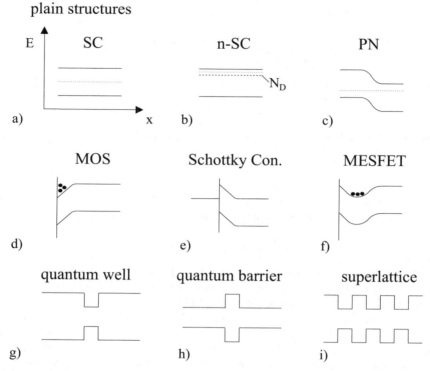

**Fig. 2.3.** Energy-band models of different semiconductor structures as they appear in devices: (a) semiconductor, (b) n-doped semiconductor, (c) pn-junction, (d) MOS structure with inversion layer, (e) Schottky contact, (f) MESFET, (g) quantum well, (h) quantum barrier, (i) superlattice

layers. For instance, quantum wells (g) as well as quantum barriers (h) can be tailored, which are of fundamental relevance for nanoelectronics. Periodic layer structures form the so-called superlattice (i). Within distinct boundaries the band structure of the superlattice can be freely modified.

Most of the mentioned plain structures appear in the next chapter. First of all the classic transistors will be analyzed in terms of their minimum dimensions.

## 2.2 Technologies

As an overview, some results of the present semiconductor technology will be presented. In this context the Semiconductor Industries Association (SIA) takes a closer look at this very active area [5].

## 2.2.1 Different Types of Transistor Integration

The minimum feature size of silicon-based transistors is an important question for the further development of microelectronics. On the basis of recent investigation results, Fig. 2.4a reveals a first-order approximation of today's integration limits. Distinct spacer layers smooth the junction between gate and drain area, which prevents an electric breakdown. Gate lengths of below $70\,nm$ and gate oxide thicknesses of about $4\,nm$ are already attainable. Such MOS transistors operate at the limits of the tunneling effect and the Debye length. They occupy less than $0.1\,\mu m^2$. These transistors can still be improved towards gate lengths of about $20\,nm$ and gate oxide thicknesses of $0.8\,nm$ [7].

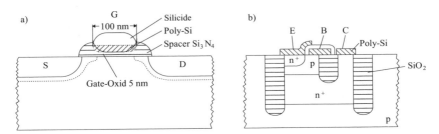

**Fig. 2.4.** Today's minimum feature sizes: (a) MOS transistor and (b) bipolar transistor.

Due to the more complex structure of the bipolar transistor, it consumes more area. Nevertheless, passive insulation and self-aligned base and emitter contacts lead to device areas of roughly $4\,nm$ (Fig. 2.4b). Passive insulations and plain surfaces are essential for future integrated circuitries. For instance, $SiO_2$ layers are ideally suited for a passive insulation. At present, such layers are uniformly deposited over the whole wafer. However, in future a local deposition via laser activation might be attainable.

The silicon on insulator (SOI) technique offers an almost ideal insulation of the devices and prevents parasitic effects. Therefore, silicon has to be deposited on an insulator, which can be achieved on different ways: (1) Oxygen has to be deeply implanted into a silicon substrate that is annealed in a second step. Thus, the $SiO_2$ layer separates the upper silicon layer from the rest of the substrate. (2) The silicon deposition on top of an insulator, e.g. sapphire, also leads to the typical SOI structure. However, this technique turned out to be too expensive. (3) It is possible to grow or to deposit a polysilicon layer on top of an $SiO_2$ layer. Afterward, the polysilicon layer is almost completely converted to a monocrystal layer by a melting procedure. The last method is very interesting when several active silicon layers are operated in a stack arrangement, which leads to a real three-dimensional integration.

Nevertheless, three-dimensional circuits become more and more important when the single-transistor area can no longer be reduced. In the past, three-dimensional integration was not an issue and higher packing densities have been realized via smaller minimum feature sizes. In general, passive insulation increases the production complexity substantially, but offers the integration of several stacked silicon layers. So-called via holes establish the interconnection of the several layers. Additionally, three-dimensional integration is very interesting for sensor systems . For instance, the sensor is integrated in the upper layer, whereas the data processing takes place in the intermediate layer and information storage is realized in the lower layer.

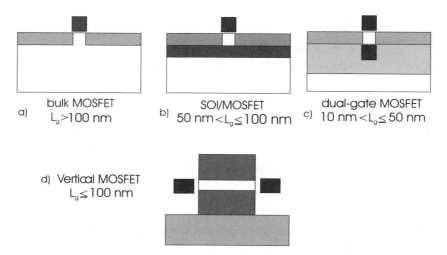

**Fig. 2.5.** Some integration-type MOS transistors with short channel lengths and small inverse leakage currents

For short-channel MOS transistors the high leakage currents present a serious problem in dynamic circuits. The cutoff current can be reduced by implementing the transistors on an insulation layer, which can be achieved by using SOI technologies. Another technique is the use of a second gate electrode, which depletes charge carriers from the channel in cutoff mode. Short channel lengths for vertical MOS transistors can also be realized by using a special series of production layers.

Figure 2.6 illustrates how fine lateral structures can already be implemented by current technologies [8]. A fine structure can be created at the edge of an auxiliary deposition layer. For example, a thin edge of polysilicon is left behind after etching polysilicon that has been deposited on an auxiliary layer. By using this spacer technique MOS transistors with a channel length

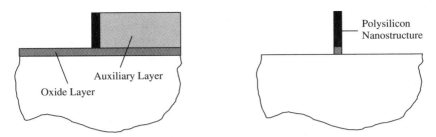

**Fig. 2.6.** Production of very fine structures using the spacer technique

down to 30 nm have already been implemented. Today these structures are utilized to study dedicated quantum effects.

### 2.2.2 Technological Processes for Microminiaturization

The concept of technological scaling represents the transition to finer structures, thinner layers, more planar surfaces, and to passive insulation. Many technological efforts are being made to accomplish these goals and it would be impossible to give an complete overview. Therefore we will present some examples to illustrate the principles of technology microminiaturization.

In solid-state microelectronics thin layers are utilized for the vertical structuring. A usual structuring process illustrated in Fig. 2.7 is molecular beam epitaxy (MBE). A heated substrate is exposed to molecular beams in a vacuum environment. These beams are emitted from so-called effusor cells. Because multiple cells can be used in parallel it is possible to achieve nearly every mixing ratio of layers and doping profile. The layer thickness is exactly determined by the opening times of the effusor shutters. Another advantage of the MBE is the possibility to produce nearly abrupt GaAs junctions. Such junctions are needed for the production of quantum-well structures. Molecular beam epitaxy can also be applied locally by masking the substrate with a $SiO_2$ layer. This method is utilized to produce resonant tunnel diodes (RTDs) and metal gate field effect transistors which will be explained in Chap. 12.

The horizontal structuring is often carried out by lithography processes, whereby short-wave radiation, for example short-wave UV, electron beams, X-radiation, and ion beams, are used to produce finer structures. Figure 2.8 presents a rough overview of the different lithography methods. At first sight it is possible to create ever finer structures by using higher-energy radiation. But it must be noted that the material defects also increase proportionately. In the near future the UV-wave exposure, using sophisticated devices, seems to be sufficient to produce feature sizes down to 80 nm.

Electron-beam direct writing is a usual method to produce fine structures. The major disadvantage of this approach is that all individual structures must be written one after the other, which consumes a lot of time. That is why

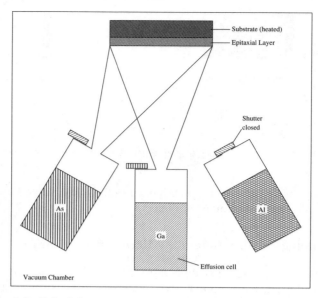

**Fig. 2.7.** Principle structure of a molecular beam epitaxy system

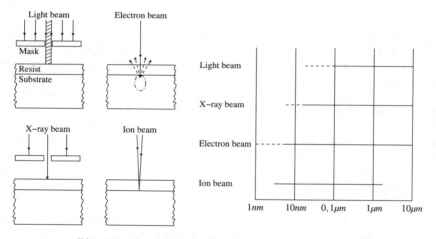

**Fig. 2.8.** Overview of different lithography methods

electron-beam writing is only economical for the mask production and not for direct structuring tasks on wafers.

The X-radiation lithography is very promising, although no usable lens systems and reflectors for wavelength between 0.5 nm and 5 nm are known. The imaging has to be performed by so-called contact copies with special masks. The mask carrier is a thin silicon film ($2\,\mu$m), which is transparent for X-radiation. The actual masking part is a thin gold layer ($1\,\mu$m) structured by electron-beam writing. In order to avoid contact copy image defects on the semiconductor wafer, the X-ray beams should run as parallel as possible. When using a normal X-ray tube, the distance between the radiation source and the silicon wafer is so small that many image defects occur in the peripheral wafer area. This error $\Delta B$ can be simply measured by the following rule:

$$\Delta B = s\frac{B}{S}, \quad (2.4)$$

where $s$ represents the mask-wafer distance, $B$ equals the wafer radius, and $S$ is the distance to the radiation source. To keep this deviation small, the distance $S$ has to be increased as much as possible, which can be achieved by extracting X-ray beams from a synchrotron. When using such a synchrotron, the distance $S$ can be chosen to be relatively large, for example 10 m, so that the error $\Delta B$ is reduced to less than $\pm 10$ nm.

Figure 2.11 presents a comparison between the different lithography methods. The highest throughput is still achieved by optical lithography. A higher resolution can be attained by using X-ray or electron beams. The single probe methods (SPMs), whereby single atoms are manipulated, yields the best results. Remarkable results are also produced by a structure printing process, the so-called nanoprinting (Fig. 2.9). Because we still encounter a lack of efficient production methods, these techniques are an important research area of nanoelectronics.

Another method should be mentioned in this context: The scanning tunneling microscope (STM) illustrated in Fig. 2.10 can be utilized to visualize and analyze the fine structures.

The principle of such a microscope is simple. A movable tip made of metal approaches a surface under vacuum conditions, until a tunnel current is recognized. The distance to the surface is controlled by this tunnel current. If the current is maintained constant, then the surface structure can be determined by measuring the tip position. In practice the distance is controlled by a piezoelectric element and the measuring tip can also be moved laterally by additional voltages, so that very small regions of a material sample can be scanned.

The STM can also be utilized to write physical structures. By taking appropriate steps, the measurement tip can pick up single atoms and reinsert them in another position (atomic force microscope, AFM). In this way complete lines or characters have been written with atoms, as can be seen in Fig. 3.8. The final objective of this technology is to deposit atoms at specified positions on a surface or in a crystal.

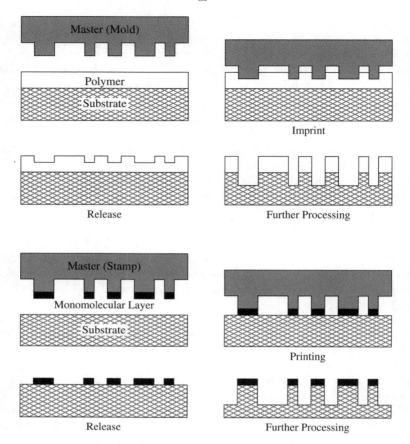

**Fig. 2.9.** Production of fine structures using printing techniques

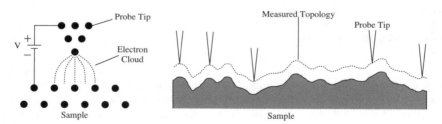

**Fig. 2.10.** Fundamental principle of a scanning tunneling microscope: A measuring tip approaches a sample until a specific tunnel current has established. The current, and hence the constant distance to the sample, is maintained by an automatic control system

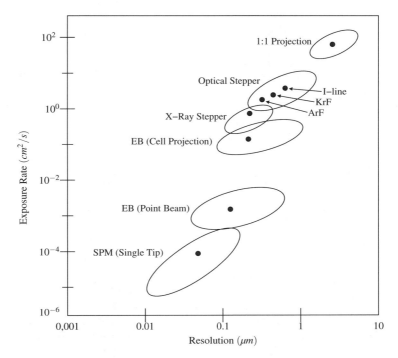

**Fig. 2.11.** Comparison of different lithography techniques from the production point of view

## 2.3 Methods and Limits of Microminiaturization in Silicon

An essential objective of microelectronics is to produce ever-smaller circuit switches. Systematic approaches to this problem were already implemented very successful in the 1970s.

### 2.3.1 Scaling

The dimensions, voltages, and currents of a switching element are decreased in such a way that the electric field strength and thus the maximum stress of that device are maintained (constant-field principle). Using this scaling method, the dimensions of a field effect transistor are decreased by a constant factor $\alpha$, as illustrated in Fig. 2.12. If this method is applied to a capacitor, then the capacity is decreased by $\alpha$:

$$C' = \frac{C}{\alpha} \qquad (2.5)$$

If this capacity has to be charged by a specific current $I$ and for a given voltage swing $V$, then the switching time $t_c$ equals

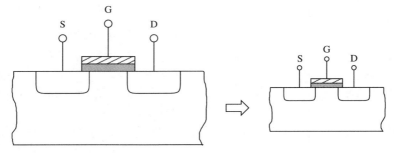

**Fig. 2.12.** Scaling: The physical dimensions and the electric quantities are scaled down by a factor $\alpha$. Thus the electric stress inside the transistor, e.g. the electric field, remain constant

$$t'_c = \frac{V'C'}{I'} = \frac{t_c}{\alpha}. \tag{2.6}$$

These considerations can generally be applied to MOS circuits. Because both the voltage $V$ and the current $I$ are scaled down by $\alpha$, the switching time $t'_c$ is also reduced by $\alpha$, which means the circuit gets faster by scaling.

The power dissipation of a circuit block is proportional to the voltage and the current, and is therefore decreased by $\alpha^2$. But because the packing density is also increased by $\alpha^2$, the unit-area power density stays constant. This result is very important in view of an integrated circuit's heat dissipation. This method is the only way to prevent heat from becoming the limiting quantity of integrated-circuit performance. In practice the supply voltages are often not reduced by the given scaling factor, so that the circuit performance is further increased, but this methodology may lead to serious heat-dissipation problems.

The power-delay characteristic is a very fundamental form of representation. Today often more than those two quantities are utilized to characterize a technology. The region 1 of Fig. 2.13 shows the characteristics of some solid-state switching elements. This category contains the devices that can be implemented today. The integrated circuits may advance in region 2, if the necessary technological research and development efforts are made. Region 3 lies beyond the classical physical limits and will not be reached, unless new concepts, for example quantum computing, are applied.

### 2.3.2 Milestones of Silicon Technology

If we compare the physical dimensions of current MOS transistors with the minimum feature sizes derived from physical limits, we will see that there is still some potential left. For these developments the Semiconductor Industries Association (SIA) has defined milestones up to the year 2015, which have been published as the so-called "roadmap 1999"[5].

## 2.3 Methods and Limits of Microminiaturization in Silicon 25

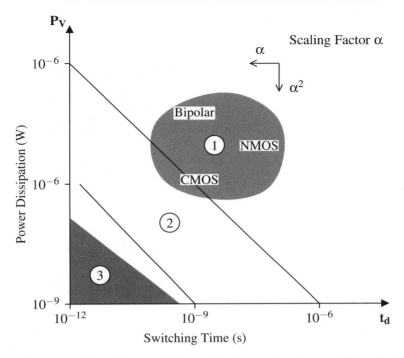

**Fig. 2.13.** Power-delay characteristics: region 1 contains the quantities of current technologies, region 2 is enclosed by technological limits, and region 3 lies beyond classical physical limits

This roadmap is updated every few years. It is interesting that the forecasted quantities often have to be corrected to higher values. A brief outline of important quantities of the very detailed 1999s plan is given in Table 2.1.

**Table 2.1.** Selected quantities of the SIA 1999 roadmap

| Year | 2000 | 2005 | 2010 | 2015 |
|---|---|---|---|---|
| Feature size [nm] | 130 | 100 | 50 | 35 |
| Device density $[\frac{10^9}{cm^2}]$ | 0.5 | 1.7 | 10 | 24 |
| Device density $[\frac{10^9}{Chip}]$ | 2 | 9 | 70 | 190 |
| Clock frequency [GHz] | 1.5 | 3.5 | 10 | 13 |

The remarkable characteristics that are predicted for the silicon technology outperform current circuits by orders of magnitude. This means that future information technology systems will also show an equivalent increase in performance [9, 10].

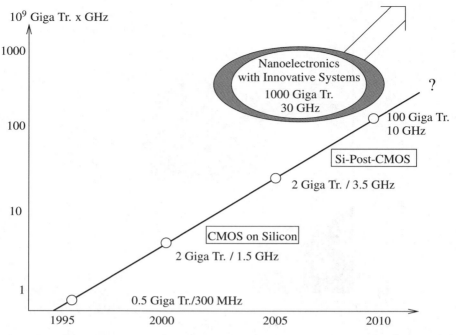

**Fig. 2.14.** Characteristic of high-performance silicon microprocessors according to the SIA studies from 1999. Nanoelectronics will only become established if they outperform classical systems by at least one order of magnitude. In the meantime the SIA corrected the given quantities to even higher values

In Fig. 2.14 the characteristics of such a development are roughly estimated from todays point of view. For this special case the product of device density and clock frequency is plotted against time. According to our assumption the characteristic of nanoelectronics should be found on top of this curve. Therefore a numerical example is also given.

The transition to nanoelectronics will only occur if it offers performance and device density increase by at least one order of magnitude, compared to standard silicon-based technologies. This transition is expected to be smooth. At first some functional blocks will be included in a hybrid architecture, and sooner or later the complete system will be implemented in nanoelectronics, whereby the peripheral devices will always consist of larger structures, because they have to communicate and adapt to our everyday environment.

### 2.3.3 Estimation of Technology Limits

In reality the method of scaling cannot be applied as ideally as we would assume from our initial considerations. For example the voltage equivalent of thermal energy $V_T$, which plays an important role in device physics, is not

## 2.3 Methods and Limits of Microminiaturization in Silicon 27

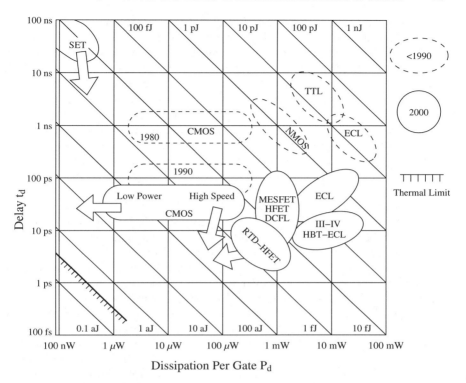

**Fig. 2.15.** Power-delay characteristic of current silicon technologies as well as more advanced SET and RTD technologies, which will be discussed later

decreased by $\alpha$, unless we are not applying lower temperatures. The current density inside the transmission line can also not be maintained constant, as the electric field, and is instead increased by $\alpha$. Hot electrons occur inside the transistors, which alter the electric properties. The RF properties of the transmission lines also worsen for smaller dimensions. These microstructure effects limit the possibilities of semiconductor electronics scaling.

When concidering circuit engineering we also have to consider the deviation of parameters from their ideal values. The variation of transistor parameters will stay constant, or rather increase after a scaling step. The variance $\Delta V_{Tn}$ of the threshold voltage will not be changed, so that the variance relative to the supply voltage will increase:

$$\frac{\Delta V'_{Tn}}{V'} = \frac{\Delta V_{Tn}}{V} \cdot \alpha. \tag{2.7}$$

This parameter deviation, which increases by downscaling of feature sizes, may lead to circuit-design problems and has to be captured by the appropriate design tools.

The physical dimensions and parameters of a minimum transistor structure can be derived from simple considerations. Based on the fact that the physical limit of microelectronics is given by the quantum-mechanical tunneling effect, the gate oxide thickness has to be at least 3 nm, to prevent disturbing tunneling currents. The oxide itself may be stressed by a maximum electric field strength of $1\frac{MV}{cm}$, if no dielectric breakdown occurs. This equals a gate voltage of 0.3 V. The maximum tolerable substrate doping that still enables an inversion layer for the given field strength, can be calculated from the MOS transistor threshold voltage equation. This results in a doping of $N_A = 10^{18}\,cm^{-3}$. That is why there is a search for alternative insulation materials, which provide higher values of $\epsilon_r$, to reduce the gate oxide field strength.

The minimum distance between the source and the drain area, which equals the minimum transistors depletion zone width, for a supply voltage of 1 V, can be estimated as 30 nm. Taking a certain safety margin into account, we should assume a minimum channel length of $L = 50$ nm.

Based upon experience the active transistor area should equal eight times the smallest surface element, which is given by 30 nm × 30 nm (the depletion zone width). So the area of minimum-sized transistor is $0.72 \times 10^4\,nm^2 = 0.72 \times 10^{-10}\,cm^2$, which corresponds to the predictions of the SIA roadmap. This means that about 4 000 minimum-sized transistors could be placed on an area that equals the crosssection of a human hair, and 14 billion transistors could be placed on 1 cm², if we neglect any kind of wiring. In the conducting state of such transistors we could locate approximately 150 electrons inside the channel zone, which would be involved in a switching event.

What about the parameter variations? In the channel zone of the volume $(50\,nm)^3$ we would find about $10^7$ silicon atoms. In spite of the very small transistor dimensions this is a remarkable number. Assuming a doping concentration of $N_A = 10^{18}\,cm^{-3}$ we could only implant 100 doping atoms. If we calculate the relative variance by (15.25), then we get approximately 10%. From the circuit design point of view this is an acceptable value.

In the circuits of today's microelectronics, charge packages of many electrons are processed. These packages are stored inside potential wells, of which the dimensions can be estimated from the Debye length (Chap. 15). Such potential wells can be found in semiconductor memories and sensors of video cameras. In the following we will estimate the number of electrons that have to be stored inside such potential wells, in order to guarantee reliable switching events.

A MOS capacitor with an inversion layer exhibits a gate-channel-capacitance of

$$C_{ox} = \epsilon_0 \epsilon_r \frac{A}{t_{ox}}, \tag{2.8}$$

where $A$ equals the capacitor area, $t_{ox}$ represents the electrode spacing, and $\epsilon_r$ is the permittivity of the gate insulator. The energy stored in the capacitor for a voltage $V_{GC}$ equals

$$W_s = \frac{1}{2} C_{GC} V_{GC}^2. \tag{2.9}$$

This energy equals the switching energy of a field effect transistor. It is remarkable that the energy is not stored in the form of a charge, but inside the electric field. The stored charge consists of $n$ electrons, so that the energy can be written as:

$$W_s = \frac{1}{2} n q V_{GC}. \tag{2.10}$$

Considering the limit of that energy, we can assume that the voltage $V_{GC}$ equals a multiple $r$ of the voltage equivalent of thermal energy $V_T = \frac{kT}{q}$, and we get:

$$W_s = \frac{1}{2} n r k T. \tag{2.11}$$

For a given operating temperature $T$ the switching energy only depends on the factor $r$ and the number $n$ of stored electrons. The minimum value of $r$ can be derived from the I-V characteristic of a pn-junction and has to be at least from 4 to 8, if the non-linearity of this characteristic is to be exploited.

In Chap. 9 a minimum switching energy of $50\,kT$ is derived from the consideration of errors caused by thermal variations. This means that at least from 12 up to 25 electrons must be involved, but these considerations are only valid for classical electronics.

A further shrinking of the physical device dimensions as well as an additional reduction of involved electrons make a quantum-mechanical modeling mandatory. In fact, microelectronics is already on a smooth transition towards nanoelectronics, as Waser shows in his book [11].

Exemplary, Table 2.2 reveals a first order approximation of the most relevant parameters of resonant tunneling devices and how they are effected when the RTD side length $L_{RTD} = \sqrt{A_{RTD}}$ is scaled by a factor of $\kappa$. The lower limit of the RTD side length is located in the range of $L_{RTD} = 20 - 40$ nm.

The gate length of the HFET is scaled separately by the factor $\gamma$. In comparison to the scaling law of CMOS the operating voltage of the MOBILE can only be slightly reduced, since the operating voltage already is less than 1.0 V. Therefore the threshold voltage of the enhancement type HFET-RTD's should not be reduced below $V_{t0} = 100$ mV. The peak voltage of the RTD as well as the operating voltage have direct influence on the voltage swing $\Delta V \approx 0.9\,V_{DD} \approx 2.5\,V_P$ and can be reduced by a factor of $\eta = 2\ldots 3$ to $V_P = 0.1\,V$. Since the logic 1-level is approximately at $V_H = 0.3V$ the drain-source current of the HFET is relatively small. This might be compensated by enlarging the gate witdh of the transistor but this also enlarges the capacitive

**Table 2.2.** Scaling law of the RTD-HFET MOBILE gate (NAND, Fan-in = Fan-out=2)

| term | definition | unit | scaling |
|---|---|---|---|
| **basic parameters** | | | |
| RTD side length | $L_{RTD}$ | $\mu$m | $1/\kappa$, $\kappa > 1$ |
| RTD-HFET gate length | $L$ | nm | $1/\gamma$, $\gamma > 1$ |
| Peak-current density | $j_P$ | kA/cm$^2$ | $\alpha$, $\alpha > 1$ |
| Peak voltage | $V_P$ | V | $1/\eta$, $3 > \eta > 1$ |
| **derived RTD-HFET param.** | | | |
| Minimum RTD area | $A_{RTD}^{min} = L_{RTD}^2$ | $\mu$m$^2$ | $1/\kappa^2$ |
| Peak current | $I_P = j_P \, A_{RTD}$ | $\mu$A | $\alpha/\kappa^2$ |
| Voltage swing | $\Delta V = 2 - 2.5 \, V_P$ | V | $1/\eta$ |
| Operating voltage | $V_{DD}, V_{CLK}^{max} = 2.5 - 3 \, V_P$ | V | $1/\eta$ |
| RTD capacitance | $C_{RTD} = \widetilde{C}_{RTD} \, A_{RTD}^{min}$ | fF | $1/\kappa^2$ |
| RTD speed-index | $SI = I_P / C_{RTD}$ | V/ns | $\alpha$ |
| RTD-HFET gate width | $W$ | $\mu$m | $1/\gamma$ |
| Gate-source capacitance | $C_{GS} = \widetilde{C}_{GS} \, W \, L$ | fF | $1/\gamma^2$ |
| **der. param. of the MOBILE** | | | |
| Load capacitance (Fan-out 2) | $C_L = C_{RTD} + 2 \, C_{GS}$ | fF | $C_{RTD}/\kappa^2 + 2 \, C_{GS}/\gamma^2$ |
| MOBILE Speed-Index | $SI_{MOB} = I_P / 2C_L$ | V/ns | $\alpha$ |
| Intrinsic MOBILE switch. time | $t_{int} = \Delta V / SI_{MOB}$ | ns | $1/(\alpha \, \eta)$ |
| Cycle | $T = 10 \, t_{int}$ | ns | $1/(\alpha \, \eta)$ |
| Power diss. during switching | $P_{sw} = 3/5 \, I_P \, V_P$ | $\mu$W | $\alpha/(\eta \, \kappa^2)$ |
| Static power diss. ($PVCR = 3$) | $P_{stat} = 2/5 \, I_P \, V_P$ | $\mu$W | $\alpha/(\eta \, \kappa^2)$ |
| Dynamic power dissipation | $P_{dyn} = C_L \, V_P^2 \, \frac{1}{T}$ | $\mu$W | $\alpha/(\eta \, \gamma^2)$ |
| Power delay product | $P_{ges} \, t_{int} \approx P_{dyn} \, t_{int}$ | fJ | $1/(\eta^2 \, \gamma^2)$ |

load of the preceding stage and therefore decreases the maximum operating frequency. In contrast to this the circuit performance and the fan-out can be controlled by the peak current density of the RTD. Therefore the scaling factor $\alpha$ has been introduced. By using different scaling factors for the HFET and the RTD a fitting of the different devices to each other becomes practical. Since HFET transconductances of about $g_m^{max} = 1000 mS/mm$ are available with gate lengths below $200 nm$, the $W/L$-ratio can be scaled down in a disproportionate fashion. For the calculation of the intrinsic switching times in Table 2.2 the simplified case $\kappa = \gamma$ has been assumed. As an important finding for the III/V technology one can state that the performance of the HFET-RTD MOBILE can be increased by only increasing the peak current density, because the lateral scaling factor $\kappa$ has no impact on the speed-index of the RTD.

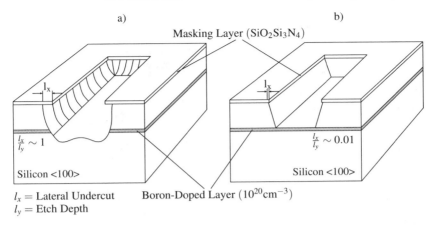

**Fig. 2.16.** Etching of structures for micromechanics with anisotripic selective etch process (b), compared to conventional etching (a)

## 2.4 Microelectronic and Mechanical Systems (MEMS)

An interesting and very promising development to more complex VLSI systems is the diversification to microelectronic and mechanical systems (MEMS).

The integration of mechanical components makes it possible to bring various elements into a system, as actuators and sensors. This will open many new areas of applications, as Fig. 2.17 indicates. MEMS will also promote nanoelectronics (NEMS), since they can provide the sophisticated interface between nanoelectronics and our environment [12].

### 2.4.1 Technology of Micromechanics

Micromechanics means the realization of mechanical structures the geometric dimensions of which are so small that they can not be realized with conventional precision work. In the past micromechanics used the outstanding characteristics of the silicon technology. Silicon belongs to the most investigated, the most understood, and technologically the most controlled materials, as mentioned in the introduction of Chap. 1. The strength of silicon is superior to steel so that tiny and complex structures can be manufactured with high yield and they operate with high reliability as well.

The realization of micromechanical parts uses a special etching technique. First, this technique applies an anisotropic selective etching that etches along the different crystal axis with different intensities. In this way we get well-shaped trenches. Secondly, the etching velocities depend on the doping of the layers, so we can realize etch stops easily. The processes use silicon nitride and silicon dioxide layers both for masking and electrical isolation.

The results of this etching technology are schematically shown in Fig. 2.16. The anisotropic selective etching yields flat walls that go along the crystal axis,

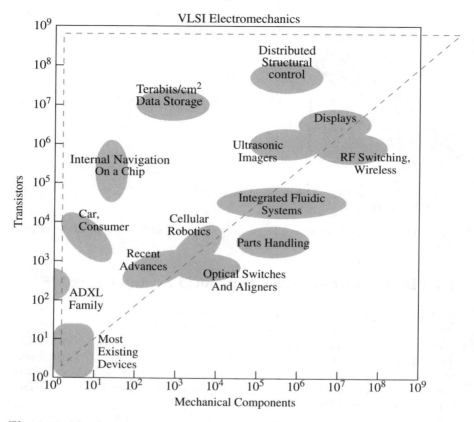

**Fig. 2.17.** Map for micro- and nanosystems on which micromechanical components are integrated together with electronics

in this example for a <100> silicon material. Since the boron-doped layer represents an etch stop we get a flat bottom in the trench. The dimensions of the trench are defined by the window of silicon dioxide at the top of the substrate.

### 2.4.2 Micromechanics for Nanoelectronics

One interesting aspect of the trenches is that they offer an integration of various functional parts into one silicon crystal together with integrated circuits. In this way different technologies can be brought together as Fig. 2.18 shows. This hybrid technology combines many various components with small structures into one chip with high accuracy. These features are the prerequisite for nanoelectronics interfaces with conventional microelectronics. This technique is in use for the connection of glass fibers to a chip in which the fiber is adjusted in a V-shaped grove so a low-loss transition from the fiber

## 2.4 Microelectronic and Mechanical Systems (MEMS)

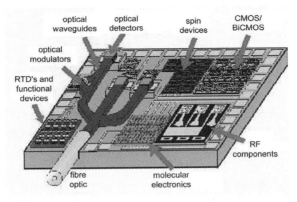

**Fig. 2.18.** Hybrid technology: Various components on one silicon chip [13]

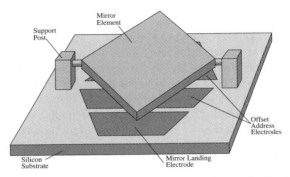

**Fig. 2.19.** An element for sweeping a light beam or for projecting light on photodiodes or a screen

to the light sensors on the chip is feasible. Such structures are also useful for high-frequency circuits.

A structure often used in micromechanics is the floating bar over a trench. The bar resists from etching since it is protected by the mask. The bar can oscillate and has a resonance frequency. Therefore we can build a frequency meter with several bars or an acceleration meter with only one bar, which can be applied in an ABS sensor of a car.

Micromechanics often is the key technology for sensors that we need in environmental technology or in the health system. A gas sensor consists of a trench with a suspended electrode. The electrode is the gate for the field effect transistor, the drain and the source of which are implemented at the bottom of the trench. The volume of the trench between the gate and the channel region can be filled up with a chemical substance that reacts with gas molecules that are to be detected. Since dielectric changes or ionized gas molecules influence the threshold voltage of the transistor the structure can measure the concentration of the gas. In the gate electrode holes can be pro-

34    2 Potentials of Silicon Technology

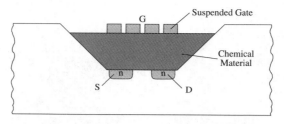

**Fig. 2.20.** Cross section of a gas sensor with a field effect transistor with a suspended gate

vided so that the gas molecules can intrude into the insulator of the trench. Such a sensor is a preceding stage of molecular electronics. In this context we should mention the microfluidic that is used in biological analysis systems on a chip, which are described in Chap. 11.

As another example Fig. 2.21 shows an interesting actuator that is movable in x, y, and z directions by applying a voltage at the piezoelectric layers and can scan a volume of about 0.001 $\mu$m$^3$. This actuator is also realized by the technology of micromechanics. The probe tip in the end of the bar can scan the surface of an object as is done at the scanning tunneling microscope. In nanoelectronics such a probe tip can write lines or program gates, and it can contact circuits, for example in memories like Millepede.

In these ways micromechanics provides essential methods for nanoelectronics.

## 2.5 Integrated Optoelectronics

In Chap. 4 it will be shown that information processing basically can be realized by using photons or electrons, and that electronics are widespread because of their well-known advantages. But photonics also offer interesting properties: Information coded by light waves can be transmitted very rapidly and can be parallel processed easily. Difficulties arise when we consider the amplification and storage of information. That is why the research results in photonics are often criticized. The following disadvantages are often quoted in this context: the lack of amplification, the unsolved fanout problem, the missing output separation from the input of circuit elements, and the difficulties concerning the implementation of bistable elements. But solutions for

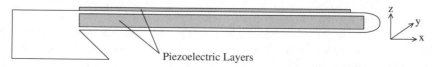

**Fig. 2.21.** Microtip for actions in the nanometer range

## 2.5 Integrated Optoelectronics

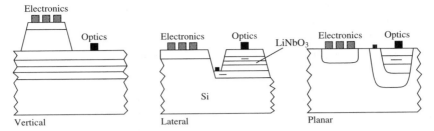

**Fig. 2.22.** Basic principles of optoelectronic integrated circuits (OEIC): vertical, lateral, and planar integration can be distinguished

these problems could be seen to emerge. A further advantage is the fact that optical information processing can be carried out with low power dissipation, in integrated circuits that occupy only a small volume.

Integrated optics are taken as miniaturized optical circuits, in which light signals are generated, conducted by waveguides, processed by corresponding physical effects, and finally detected by sensors. Special circuit components for these functional elements have been developed. It is attempted to integrate each of these components monolithically on a single substrate. The subsequent step is the attempt to integrate photonic and electronic circuit elements on a single chip. These kinds of systems are called optoelectronic integrated circuits (OEICs). Different technologies for the implementation of OEICs are illustrated in Fig. 2.22.

GaAs lasers are often used as light sources of integrated optics (Fig. 2.23). Because of their small physical dimensions, microlasers are useful components of nanoelectronics. Today, semiconductor lasers cannot be implemented in silicon because this material is characterized by an indirect bandgap. By utilizing micromechanics the silicon substrate can be prepared for a hybrid technique for mounting of a laser diode. Another approach is given by heteroepitaxy, whereby GaAs layers are deposited on a silicon surface. There is also work progressing to determine if impurity atoms can be inserted in a silicon crystal in such a way, that a direct band transition of electrons becomes possible. The first promising results, concerning the insertion of carbon atoms, have been reported.

GaAs light-emitting diodes can be used as incoherent light sources. The application of light-emitting silicon is also being examined, although the efficiency factor of that material, as known today, is very low.

As a first example, an AND gate integrated on a GaAs chip is presented in Fig. 2.24a. Two phototransistors, which control a photodiode, are connected in series: The photodiode only emits light if an optical signal is applied to both inputs $A$ and $B$. Thereby the input signal is amplified. The schematic of Fig. 2.24b illustrates the cross section of an integrated device: the stacked layers of the phototransistors and the photodiode lie on top of each other.

Integrated optical waveguides, which are used to connect integrated optical circuit devices, consist of a layer, with the refractive index $\eta_1$, conducting the wave, and of further surrounding layers with a refractive index $\eta_2$ smaller than $\eta_1$. Because of the physical dimensions and the refractive index profile, only specific wave modes will be carried by the waveguide.

In conjunction with standard silicon technology, integrated optical waveguides can be realized by a small extra effort, for example as ribbon waveguides, or as waveguides buried in the substrate. In the following we will focus on waveguide layers made of oxides, which can be easily produced in an extended CMOS technology (Fig. 2.25).

Photonics are particularly interesting for complex connected structures, because they enable a high degree of parallel processing, because light waves have no need for electric conduction, and because they offer a high immunity to electromagnetic disturbance. Light waves are characterized by very high frequencies, so that they are very suitable for bulk transmission of data. If we consider an optical fiber, we have a bandwidth of 50 000 GHz at our disposal.

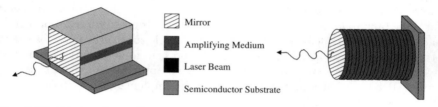

**Fig. 2.23.** Semiconductor laser diode and a microlaser. The physical dimensions of the microlaser are $7\,\mu$m diameter, and $10\,\mu$m length

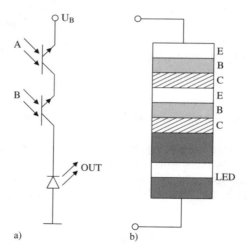

**Fig. 2.24.** Nonlinear elements for optoelectronics: (a) an AND gate consisting of phototransitors and a photodiode, and (b) its technological implementation

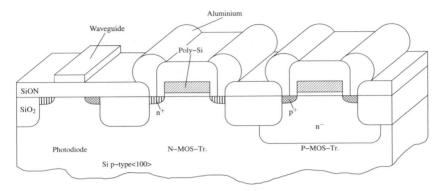

**Fig. 2.25.** Optical integrated circuit in CMOS technology

With this huge bandwidth we can feed the inputs of very complex nanoelectronic chips by way of example. This can be achieved by using parallel light rays, or by using a single light wave and a particular modulation procedure.

An interesting application for nanoelectronics combined with nanotechnology provides the intelligent dust, shown in Fig. 2.26. Such a dust particle consists of a complete data-processing system. In this case nanostructures are mandatory. In addition, the system combines optoelectronics and power supplies, for example with a thick-film battery and a solar cell. Intelligent dust can help to solve difficult problems such as the efficient movement of the fingers of a piano player or the optimization of a jet stream. It is an example for ubiquitous computing.

## 2.6 Conclusion

Today, silicon is still the most important base material for microelectronics, because it is very suitable for integrated circuits and thereby covered by in-depth research. By using special configurations of the silicon material, or by adding other materials, the limits of the silicon-based circuits have been extended ever further. The transition to higher component densities and finer physical structures demands new technological steps, which have to treat the involved materials very carefully, and that parameters have to vary as little as possible.

Today we know a significant number of technological limits, which seem to be almost insuperable. We have to assume that this limits can be extended by spending huge scientific efforts. The main research goal in this area is not only to avoid the parasitic effect that will occur in the course of this development, but also to directly use this effect in circuits and systems.

From an electronic point of view, it is at first unexpected that micromechanical components can be produced in silicon. But this approach also shows

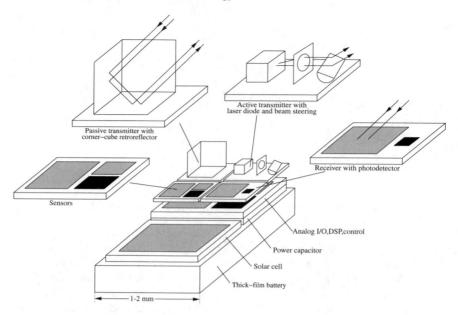

**Fig. 2.26.** Intelligent dust and its components

that an unusual course of thinking may lead to interesting innovations. The research topic of sensors and actuators will gain economic importance in the near future, for example as components in complex systems for environmental monitoring. To realize such devices, it will be necessary to implement various physical and chemical effects in silicon chips, in order to enable integration-related system production.

# 3
# Basics of Nanoelectronics

For an introduction to nanoelectronics it will be necessary to use quantum-mechanical concepts to explain the operational principles of these very fine structured devices. In this chapter the basic principles of quantum theory will be presented from a conceptual point of view. Quantum effects set up the limit of electronic miniaturization and information processing. They will be of great importance, if device dimensions are of the order of the de Broglie wavelength [14]. Another kind of quantization may be relevant to information processing. Therefore some principles of information theory will be presented at the end of this chapter.

## 3.1 Some Physical Fundamentals

### 3.1.1 Electromagnetic Fields and Photons

In electrical engineering, electric and magnetic fields, voltages, and currents occur. These values as well as electromagnetic waves are described by Maxwell's equations:

$$rot H = J + \frac{\partial D}{\partial t} \qquad (3.1)$$

$$rot E = -\frac{\partial B}{\partial t}. \qquad (3.2)$$

$H$ is the magnetic, $E$ the electric field strength, $J$ corresponds to the current density, $D$ represents the dielectric and $B$ the magnetic flux. The electric and magnetic fields in physical structures, e.g. in a waveguide, can be calculated by using these partial differential equations with the correct boundary conditions. Because these equations describe macroscopic electrical effects, they represent the basics of classical electrical engineering. The mesh and the nodal rule that are used to calculate the voltages $V$ and currents $I$ in an electrial network derive from these equations.

## 3 Basics of Nanoelectronics

The electric current $I$ is caused by the motion of charge carriers. Such carriers in semiconductors are electrons or holes. In microelectronic switching elements, e.g. in transistors, electric fields control the current.

An interesting nanoelectronic problem is the storage of a charge package that is as small as possible on a very small capacitor. Since the amount of charge $Q$ is quantized, it consists of $n$ elementary charges $q$. From Coulomb's law we can write the energy stored in the capacitor as

$$W = \frac{1}{2}\frac{(nq)^2}{C}. \tag{3.3}$$

Thus the energy $W$ on the capacitor is increased by minimizing the capacity $C$. This effect can be reached by reducing the capacitor dimensions to the nanometer scale. If additional elementary charges $q$ are applied to the capacitor by using a tunnel element, the stored energy of the capacitor increases abruptly stepwise. The so-called Coulomb blockade is based on this effect. Assuming that the potential of a very small capacitor is below the potential of a charge source, an electron can only flow from source to the capacitor if the potential on the capacitor after the transition is still below the source potential. Otherwise the process would not obey the principle of energy balance. The single-electron transistor uses this effect, which will be introduced in Chap. 13.

In order to calculate the behavior of a wave, for example the electric field distribution $E(x,t)$ on a conduction line, we can derive the following differential equation from Maxwell's laws:

$$\frac{\partial^2 E}{\partial x^2} = \frac{1}{c^2}\frac{\partial^2 E}{\partial t^2}. \tag{3.4}$$

The electromagnetic fields of light beams (Fig. 3.1) are described by this equation as well. The energy flow is characterized by the Poynting vector, which follows from the field distributions. In this case the constant $c$ represents the speed of light and the wave impedance of vacuum is given by $Z_0 = \frac{1}{c\epsilon_0} = 377\,\Omega$.

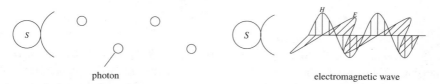

Fig. 3.1. Light as a particle flow or an electromagnetic wave

Light can alternatively be regarded as a flow of energy packets. These particles are called photons, both the wave and the particle model coexist.

This dualism is expressed by Einstein's hypothesis that associates the photon energy with the frequency of the light wave:

$$W = h \cdot f = \frac{hc}{\lambda}, \tag{3.5}$$

where $h$ is the elementary quantum of action (Planck's constant). By increasing the frequency $f$ of a light beam, the energy packets $W$ of the photons become larger. If the radiation intensity, i.e. the wave amplitude is increased, then the number of photons increases likewise. The light wavelength $\lambda$ is inversely proportional to the light frequency $f$. Concerning the visible light spectrum, the frequency $f$ is in the range of a tenth of a micrometer. Quantum electrodynamics demonstrate the particle model's use. Feynman shows that all physical effects described by the wave theory can also be explained by a particle model.

By nature light is suitable to transmit information. By the actual state of knowledge of the speed of light is the fastest possible propagation velocity. But the information-processing abilities of light are restricted, because until now no switches are known that control light as well as transistors control the electron currents.

### 3.1.2 Quantization of Action, Charge, and Flux

Classical physical models assume the continuity of quantities and involve no restrictions concerning very small physical structures. The quantum theory shows, however, that values of some measurable variables of a system, can attain only certain discrete values. The smallest possible jumps in the values of those observables are called "quanta". A quantum is the smallest possible unit that has to fit into a nanoelectronic structure. The photons or lightquanta were already mentioned above [15].

Another physical quantity that does not occur as a continuous variable is the value of action $H$, which is defined as the product of energy and time,

$$H = W \cdot t = n_H \cdot h, \tag{3.6}$$

thus action is made up of $n_H$ action quanta $h$. Common electrical values are the energy $W = \frac{H}{t} = VIt$ and the power $P = \frac{W}{t} = VI$. Other quantized values may be derived heuristically:

1. The product $I \cdot t$ corresponds to the charge $Q$, which is quantized by charge packages of $q$. This elementary charge $q$ is the smallest possible charge quantity. From quantum electrodynamics we can derive the following relationship for the elementary charge

$$\begin{aligned} q^2 &= 2\alpha hc\epsilon_0 \\ q &= \sqrt{2\alpha c\epsilon_0} \cdot \sqrt{h} = 1.60 \times 10^{-19} \text{ A s}, \end{aligned} \tag{3.7}$$

where $\alpha$ is the fine structure constant given approximately by $\frac{1}{137}$. The elementary charge is proportional to the square root of the action quanta $h$. The Coulomb-blockade effect, which was introduced above, is based on the quantization of charge.

2. The product $Ut$ represents the magnetic flux, which is also quantized. If we consider the magnetic flux in a current-carrying superconductive ring as a function of the current, we notice that this flux changes abruptly stepwise. It consists of flux-quanta given by

$$\Phi_0 = \frac{h}{2q}$$

$$\Phi_0 = \frac{1}{2\sqrt{2\alpha c \epsilon_0}} \sqrt{h} = 2.07 \times 10^{-15} \text{ V s}. \tag{3.8}$$

This relationship derives from (3.6) for $n_H = 1$ and from the fact that two electrons are always contributing a supercurrent. The flux quanta $\Phi_0$ is an elementary quantity comparable to the elementary charge $q$ and is likewise proportional to the square root of the action quanta $h$. If we also take the wave impedance of vacuum $Z_0 = 377\Omega$ into account we find the following interesting relationship

$$\frac{q}{\Phi_0} = \frac{4\alpha}{Z_0}. \tag{3.9}$$

This equation uncovers the relation between the flux quanta, the wave impedance, and the elementary charge. It will be useful for the comparison of field effect electronics with superconducting electronics, which will be drawn in Chap. 14. We will see that the superconducting electronics are not well suited for device miniaturization because the action due to the flux quanta is much higher than that due to the elementary charge.

### 3.1.3 Electrons Behaving as Waves (Schrödinger Equation)

The model of the dualism of light can be transferred to other particles, e.g. electrons, because a physical mass can be associated with the photons energy, according to the relation $E = mc^2$. An electron of mass $m_e$ has a negative charge of $e = -q$. Both of these values can be measured experimentally [16].

From a heuristic point of view we can conclude the following: If the electron is in motion it shows a kinetic energy of $E = \frac{1}{2}mv^2$. Starting from the classical model such an electron would travel along with velocity $v$ and momentum $p = mv$. (As we focus on the one-dimensional case we can use scalar values instead of vector notation.) The action of this process is given by

$$H = mv \cdot x, \tag{3.10}$$

which should increase continuously with $x$. But because action is quantized, it increases in abrupt steps, which results in an abruptly changing velocity $v$ and

## 3.1 Some Physical Fundamentals

also in sudden changes of the electron energy. This contradiction is resolved by giving up the classical model of a continuous trajectory and moving on to quantum mechanics. According to this theory the electron is described by successive quantum-mechanical states, which represent a certain probability that the particle may be located in a specific spatial region.

These measures of probability can be calculated from the wave function $\Psi$ that results from the solutions of a partial differential equation called the Schrödinger equation

$$-\frac{\hbar^2}{2m}\frac{\partial^2 \Psi(x)}{\partial x^2} + V(x) \cdot \Psi(x) = E \cdot \Psi(x), \qquad (3.11)$$

where $V(x)$ is a given potential function and $\hbar = \frac{h}{2\pi}$. Considering that $V(x)$ is the Coulomb potential of a charged nucleus, e.g. of a hydrogen atom, we can calculate the wavefunctions $\Psi$, also called states, by solving (3.11). This wave representation describes the behavior of the hydrogen atom more appropriately than the well-known model postulated by Bohr in the 1930s, which involves electrons orbiting the nucleus.

If the boundary conditions change in time we have to apply the time-dependent Schrödinger equation

$$-\frac{\hbar^2}{2m}\frac{\partial^2 \Psi(x,t)}{\partial x^2} + V(x,t) \cdot \Psi(x,t) = j\hbar \frac{\partial \Psi(x,t)}{\partial t}. \qquad (3.12)$$

The Schrödinger equation is not identical to Maxwell's equations, although it is treated with a very similar mathematical formalism, so engineers should have no problemes in applying this equation.

According to the simple case of $V(x) = 0$ (3.12) is solved by the complex wavefunction

$$\Psi = A \cdot \exp\left(j[kx - \omega t]\right). \qquad (3.13)$$

The wave vector $k$ is often used instead of the wavelength $\lambda$ to characterize waves. In contrast to the treatment of alternating currents, where complex syntax is only used as an elegant mathematical methodology, the complex representation is inevitable in this case. The probability $P$ of finding the particle in a specific spatial region is defined by $\Psi^* \cdot \Psi$, or $|\Psi|^2$, respectively. If we are considering the three-dimensional case we have to perform an integration over the volume $V$

$$P = C \cdot \int_V \Psi^* \Psi dV. \qquad (3.14)$$

This equation reduces to a line integral over $x$ for the one-dimensional case. The normalization constant $C$ has to ensure that the integral over the entire possible space results in $P = 1$, because the probability of finding the particle somewhere in this region must be unity.

This interpretation of $|\Psi|^2$ suggests the introduction the term information. The information delivered by a measuring process is inversely proportional to the probability of localizing a particle in the observation space. Although this obvious relation to information theory is interesting, the concept was not generally adopted by physicists.

At the present time a direct derivation of Schrödinger's equation is not known. A heuristic one can be deduced from the observation of a particles quantized impulse according to its location $x$. This relation is represented by a step function where the probability density $\Psi$ at the discontinuity points must vanish. This condition is satisfied by applying a wavefunction like Schrödinger's equation. By using this approach we solve the problem of impulse quantization but also give up the classical theory of mechanics, where a particle moves along a continuous trajectory, as mentioned above.

The wavelength of $\Psi$ follows directly from the quantization if we assume that $H = h$ and $x = \lambda$ in (3.10):

$$\lambda = \frac{h}{m \cdot v},$$

and for the particle momentum

$$p = \frac{h}{\lambda}. \tag{3.15}$$

The wavelength $\lambda$ depends on the propagation velocity $v$, and hence on the frequency $f$, which is often written as $\nu$. The higher the momentum $p$ or the kinetic energy of the particle, the shorter the wavelength $\lambda$ of the material wave. The kinetic energy of the particle is given by

$$E = \frac{h^2}{2m}\frac{1}{\lambda^2} = \frac{\hbar^2}{2m}k^2. \tag{3.16}$$

For an energy value of 1 eV, which is typical for switching elements, the characteristic wavelength is 1.2 nm. The dimensions of nanoelectronic devices are in the same order of magnitude.

Another important relation that derives heuristically from the model described above is Heisenberg's uncertainty principle: The higher the momentum $p$, the shorter the wavelength $\lambda$, because from (3.15) we get

$$\frac{\lambda}{2} \cdot mv = \frac{h}{2}. \tag{3.17}$$

This relation is also valid for differences of impulse and location. Assuming that $\frac{\lambda}{2} = \Delta x$ and $m \cdot \Delta v = \Delta p$ we obtain the uncertainty relation:

$$\Delta x \cdot \Delta p \geq \frac{h}{2}. \tag{3.18}$$

By extending the expression on the left side we get the alternative formulation

$$\Delta E \cdot \Delta t \geq \frac{h}{2}. \tag{3.19}$$

The uncertainty principle denotes that the location or the momentum of a particle, and its energy or its time of observation can only be determined imprecisely. This statement is very important if we are considering nanoelectronic applications, because the dimensions of such devices are so small that we can use the uncertainty principle to roughly estimate the relevant nanoelectronic effects, for example the tunneling effect.

In the following sections some important nanoelectronic structures will be discussed. Thereby it is inevitable to apply the wave model of matter to describe the behavior of the electrons involved. The upcoming example of the potential well shows that it is not possible to correctly determine the behavior of an electron in such a configuration by using the classical-particle model.

### 3.1.4 Electrons in Potential Wells

Figure 3.2 depicts the simple case of a one-dimensional potential well of length $L$ that is enclosed by infinite potential walls. Because the electron is freely moving inside the well, its energy is given by

$$E = \frac{\hbar^2}{2m} k^2. \tag{3.20}$$

However, the particle is reflected at the barrier positions $x = 0$ and $x = L$, thus stationary waves build up inside the well. A standing wave can only exist if the length $L$ of the potential well equals a multiple of $\frac{\lambda}{2}$. All other possible wave instances will vanish due to destructive interferences. The stationary waves are described by the solutions of Schrödinger's equation

$$\Psi = \left(\frac{2}{L}\right) \sin \frac{n\pi x}{L} \quad \text{with} \quad n = 0, 1, 2 \ldots. \tag{3.21}$$

This solution shows that an electron can never be located at the wall, because the wavefunctions always vanish at these specific locations. This condition results in the fact that the particle can only assume discrete energy values

$$E_n = \frac{n^2 h^2}{8nL^2} = \frac{1}{2} \frac{h}{2\pi} n\omega_1. \tag{3.22}$$

This result is generally represented by a so-called band diagram, or energy diagram $W = f(k)$. In this special case the energy band mentioned above consists only of discrete values. The number of these energy values can be determined by the length $L$. The smaller the length of the potential well, the higher the energy $W_n$, which can also be written as a function of the factor $n$ and the fundamental oscillation frequency $\omega_1$. The discretization of energy values plays an important role in nanoelectronics and should not be confused with the quantization of charge that leads to the Coulomb blockade described above.

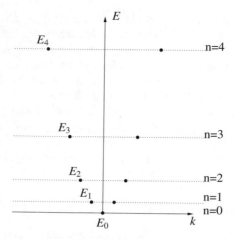

**Fig. 3.2.** The discrete energy states of an electron in a potential well. The eigenvalues lie on a parabolic curve

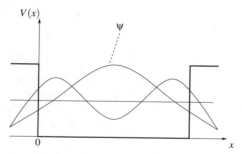

**Fig. 3.3.** A potential well, an important element of nanoelectronics. Illustrated is the 1st and 3rd solution of Schrödinger's equation

A fundamental structure of nanoelectronics is the potential well enclosed by finite walls. Because of the complex boundary conditions we obtain a slightly different solution, which is qualitatively presented in Fig. 3.3 [17].

An essential aspect of this structure is the fact that the wave amplitude inside the potential barrier is not vanishing but exponentially decreasing. The same effect can be found by observing quantum-mechanical tunneling, which plays an important role in nanoelectronic switching elements, e.g. tunnel diodes.

### 3.1.5 Photons interacting with Electrons in Solids

The fundamental physical effect of most optoelectronic devices is the interaction between photons and electrons in solid-state elements. Thereby energy is taken up by absorbing a photon and generating a pair of charge carriers,

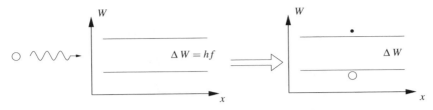

**Fig. 3.4.** Energy-band model illustrating the interaction of electrons and photons in crystal structures

namely an electron and a hole. By looking at the energy-band model we can illustrate this process by an electron that is elevated from the valence band to the conduction band (Fig. 3.4). The energy that is consumed for this transition derives from (3.5). The action converted by the switching operation is given by

$$W_s t_s = nh, \qquad (3.23)$$

which is proportional to the number of quanta. If we consider the case of $n = 1$ this leads to the minimum amount of energy that must occur at least at every switching process:

$$W_s \geq \frac{h}{t_s}. \qquad (3.24)$$

If we substitute the switching time $t_s$ by the reciprocal of the frequency $f$ we obtain the well-known relation between energy and frequency (3.5). Because the wavelength $\lambda$ is inversely proportional to the frequency $f$, the switching frequency is raised if we reduce the wavelength and thus the switching energy is increased. This is important information that shows that the switching energy is not continuously decreasing, but increasing if structure sizes are reduced to the quantum-mechanical region. Based on this model we can derive a quantum-mechanical limit for nanoelectronic devices, which will be discussed later.

The working principle of photodiodes, which are among others used as information receivers, can be explained by the interaction described above. A photodiode receives an electromagnetic wave and converts it into an electrical signal (Fig. 3.5). This principle is also used in molecular electronics, because molecules can usually be activated only by light and not by electrical signals, like transistors are.

From this model we can derive the so-called quantum noise. Photons do not arrive uniformly at a receiver, but follow a statistical distribution. Thus the useful signal is overlaid by noise, which average power $\bar{P}$ for the given frequency $f$ and bandwidth $\Delta f$ can be written as

$$\bar{P} = \frac{1}{2} h f \Delta f. \qquad (3.25)$$

The noise voltage drop at a resistor $R$ follows from the theory of electric quadrupole:

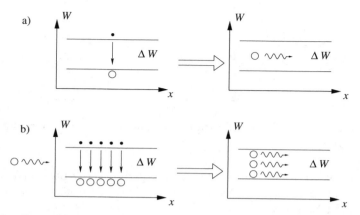

**Fig. 3.5.** Generation of photons: (a) emission and (b) induced emission of light

$$\bar{v}^2 = 4R\bar{P} = 2Rhf\Delta f. \qquad (3.26)$$

The quantum noise is the stronger, the higher the frequency and the bandwidth. It occurs particularly in systems operated at high frequencies and plays an important role in high-performance integrated circuits. The exact quantitative derivation follows from thermodynamics by considering the power of a blackbody radiator and an harmonic oscillator.

Another important interaction effect between photons and electrons is the opposite of charge-carrier generation: An electron can drop back from the conduction band to the valence band. Thereby it releases energy and emits a photon. This process is illustrated in Fig. 3.5 and occurs, for instance, in light-emitting diodes (LEDs). Another important effect is the so-called stimulated transition. If light radiates onto a current-carrying semiconductor with many electrons in its valence band, additional transitions are induced by equiphasal waves, which amplifies the incident light. This principle represents the basic operation of laser diodes.

The efficient emission of light is only possible for semiconductor materials with direct band transitions, for example GaAs. The band transition of silicon is indirect, thus it is not possible to induce an efficient emission of light. An alternative approach is the use of bremsstrahlung: Electrons in the channel of MOS transistors are accelerated to high velocities and afterward slowed down abruptly by positively charged donor atoms in the drain region. The kinetic energy released by this sudden change of velocity is emitted as radiance. Because the efficiency of this process is very low it is not implemented in technical applications, but it is used in the quality assessment of MOS transistors.

### 3.1.6 Diffusion Processes

First we will examine how the total energy of a system is distributed among its particles in the state of thermodynamic equilibrium. Therefore we consider the energy distribution of electrons in a crystal. This energy is based on heat and its average is proportional to the temperature T:

$$W = \frac{1}{2}kT, \tag{3.27}$$

where $k = 1.38 \times 10^{-23}$ J K$^{-1}$ is the Boltzmann constant.

To determine the probability that an electron takes a specific energy value we first have to calculate the entropy of an electron gas. Then we choose the distribution function characterized by a maximum entropy value. This methodology leads to the so-called Fermi'Dirac distribution

$$f_F = \frac{1}{1 + \exp\left(\frac{W - W_F}{kT}\right)}. \tag{3.28}$$

This results in a rectangular function if we choose $T = 0$. The probability $f_F$ equals one for every energy below the Fermi energy $W_F$, i.e. all energy states are occupied. On the other hand $f_F$ vanishes for energy values above $W_F$, this means no states are occupied. If we consider high energy values the Fermi distribution can be replaced by the Boltzmann statistic from (3.30). Manipulating the terms (3.28) becomes

$$\frac{1}{f_{Fi}} - 1 = \exp\left(\frac{W_i}{kT} - \frac{W_i}{kT}\right) = C_F \cdot \exp\left(\frac{W_i}{kT}\right), \tag{3.29}$$

where we transformed the continuous representation to a discrete function.

By manipulation of terms on the left side of this equation we find a distribution function for other types of particles. The Boltzmann distribution function for a gas of atoms or molecules is given by

$$\frac{1}{f_{Bi}} = C_B \cdot \exp\left(\frac{W_i}{kT}\right), \tag{3.30}$$

and by inverting the sign before the 1 in (3.29) we get the Bose'Einstein distribution for a photon gas

$$\frac{1}{f_{BEi}} + 1 = C_E \cdot \exp\left(\frac{W_i}{kT}\right). \tag{3.31}$$

This equation's modifications derive from the mode of particle distribution in the phase space.

If the density or energy of particles inside a crystal is nonuniformly distributed, a diffusion process occurs, which attempts to restore the state of equilibrium. These processes are important for nano- and microelectronics, for example in the case of charge transport, but they are not always desired, for instance when physical structures in real devices are changing over time.

In the following we will develop a plausible approach to the diffusion process by starting from the simple model in Fig. 3.7. We assume that the density

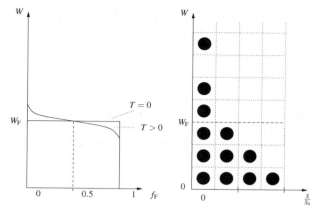

**Fig. 3.6.** Fermi distribution for $T = 0\,K$ and $T > 0\,K$. $W_F$ is the so-called Fermi level, where the slope of the function equals $kT$. A simplified model of the distribution process is presented on the right side of this diagram

of electrically neutral particles is high on one side of the crystal and low on the other. Because of the thermal movement the total flow rate between a region with a high particle density and a region with low density is proportional to the gradient of the density distribution. Thus the following differential equation describes the diffusion process:

$$S_D = D \cdot \left(-\frac{\partial n}{\partial x}\right), \tag{3.32}$$

where $S_D$ is the particle flow, $D$ the diffusion constant, and $\frac{\partial n}{\partial x}$ represents the density distribution gradient. To gain a better understanding we should take a closer look at the diffusion constant $D$. Therefore we start from a simple model of the diffusion process. We assume that the potential wells depicted in Fig. 3.7 are separated from each other by potential barriers of height $W_A$. The particles can only be transmitted to a neighboring well if they get past the barrier. The number of particles that carry the energy needed for this transition is given by the Boltzmann factor from (3.27). Thus we obtain the particle flows in both directions:

$$S_{D1} = c\,a \cdot n(x) \cdot \exp\left(-\frac{W_A}{kT}\right) \quad \text{and} \tag{3.33}$$

$$S_{D2} = c\,a \cdot n(x+a) \cdot \exp\left(-\frac{W_A}{kT}\right), \tag{3.34}$$

where $c$ is a constant factor, $a$ equals the distance of the potential wells, $a \cdot A$ is the volume of a well, $n$ represents the particle density, $a \cdot A \cdot n(x)$ determines the number of particles in the observed region ($A$ is the cross-sectional area and is set to one), and $\exp\left(-\frac{W_A}{kT}\right)$ equals the Boltzmann factor. The resulting particle flow can be written as

$$S_D = S_{D1} - S_{D2} = c\,a \cdot (n(x) - n(x+a)) \cdot \exp\left(-\frac{W_A}{kT}\right), \quad (3.35)$$

and by using differential notation we get

$$S_D = -c\,a^2 \cdot \frac{\partial n}{\partial x} \cdot \exp\left(-\frac{W_A}{kT}\right). \quad (3.36)$$

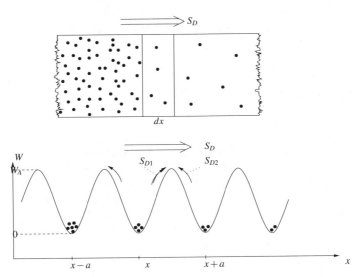

**Fig. 3.7.** Model for explaining the diffusion process and for the derivation of the diffusion constant

If we compare this equation with the particle flow in (3.32), the diffusion constant can be written as

$$D = c\,a^2 \cdot \exp\left(-\frac{W_A}{kT}\right). \quad (3.37)$$

The diffusion constant, and the diffusion current decrease with higher potential barriers and lower temperatures.

If we consider the continuity relation, which describes the particle-density variation with time in dependence from the local variation of the particle flow density,

$$\frac{\partial n}{\partial t} = -\frac{\partial S_D}{\partial x}, \quad (3.38)$$

the important diffusion equation can finally be written as

$$\frac{\partial n}{\partial t} = D \cdot \frac{\partial^2 n}{\partial x^2}. \quad (3.39)$$

This partial differential equation represents the basic law for all physical diffusion processes. Even thermal conduction can be explained by a diffusion process. The thermal current, i.e. the dissipated power, is proportional to the temperature gradient and follows from the diffusion equation (3.32)

$$P_{th} = -\kappa \frac{\partial T}{\partial x}, \qquad (3.40)$$

where the parameter $\kappa$ represents the specific thermal conductivity. The thermal current through a circuit, i.e. the power dissipated by a chip, can be calculated with this relation. Device limits caused by this effect will be discussed in Chap. 15.

## 3.2 Basics of Information Theory

### 3.2.1 Data and Bits

The beginnings of information theory can be attributed to telecommunications. At first, the goal of information theory was to efficiently transmit information (messages), e.g. along a wire line. The information content of such messages is higher the more improbable their occurrence. It was also shown that all information can be coded in the binary digits 0 and 1.

In the following we will deal with the question: What is information? If information should be processed in technical systems, it must be represented in some kind of physical structure. This mapping onto hardware considerably limits information-processing capabilities. That is why these considerations are essential for micro- and nanoelectronic systems [18].

To introduce the subject we will compare an information carrier of the past with one of the future: on the one hand the punched paper tape, and on the other hand single atoms deposited on a substrate surface. In the first case combinations of holes are punched in a strip of paper, and in the second example arrangement of atoms form letter symbols (Fig. 3.8). The symbols coded onto one of these information carriers can be read by one single person or can be viewed simultaneously by a group of individuals using some kind of projection. The combinations of holes or atoms form data, which can be processed or transmitted. The recipient of that data will gain information and is enabled to extend his stored data volume.

Examining this simple case we can derive the four important characteristics of information by U. Baitinger:

1. Information or data can be transmitted and processed without knowing its specific meaning.
2. Some kind of carrier is needed for information storage: matter or energy. In the example above paper is used to store the data. Wether information can exist without a carrier is a question of philosophy of nature.

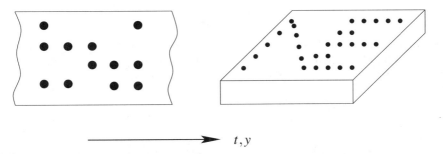

$t, y$

**Fig. 3.8.** Schematic picture of a punched paper tape, where the $y$ coordinate is scanned over time $t$, and a silicon substrate with single atoms applied to its surface by a scanning tunnel microscope (STM). Both samples represent information

3. The information carrier can be exchanged without losing any information. For example, the data can be written from a punched paper tape to a semiconductor memory.
4. The transmission and processing of information can be distorted. To transmit and process information without any loss is the challenge for information technology.

According to the third rule we assume that the information carrier can be neglected and that abstract information can be treated in forms of mathematical formalism.

In technical systems, data are processed and transmitted. For example, we can observe series of data on a line. Because we generally do not know the upcoming data word, its arrival is characterized by some means of surprise. This kind of surprise can also be described as information, which is determined quantitatively by the probability that a specific pattern $X_r$ occurs. The measure of surprise is the greater, the more seldom a pattern occurs. Therefore the information of a pattern $X_r$ is inversely proportional to the probability of its occurrence $P_r$. According to Shannon the definition of the information content $I_r$ of a pattern $X_r$ is written as

$$I_r = -K \ln P_r, \tag{3.41}$$

where $K$ is a constant factor. The logarithmic function is chosen because it allows two information content measures to be easily added and because large values of $I_r$ are scaled down to a smaller range.

In information theory, information content is measured in bits. This unit is defined by a so-called dual step, where the two patterns 0 and 1 arrive with the probability $P_r = 0.5$. Thus the factor $K$ is determined according to

$$I = -\text{ld}\, 0.5 = 1\,\text{bit}, \tag{3.42}$$

and the information content of a pattern $X_r$ can be written as

$$I_r = -\operatorname{ld} P_r. \tag{3.43}$$

If the pattern $X_r$ is chosen from a set of $m$ patterns with an identical occurrence probability, then the average information content equals $H$, also called expectation value or 'following thermodynamics' entropy.

$$H = -\sum_{i=1}^{m} P_i \cdot \operatorname{ld} P_i \qquad i = 1 \ldots r \ldots m. \tag{3.44}$$

In the following we are considering the special case of $m$ different incidents with the same occurrence probability $P_r = \frac{1}{m}$. The average information content reaches a maximum value given by

$$H_0 = -\operatorname{ld} P_r = \operatorname{ld} m. \tag{3.45}$$

We can now examine patterns for many different applications and calculate their information content. For example, a bus system of 6 signal lines, where one line is always in an active state, could be compared to a game of dice (Fig. 3.9). This signals could trigger switching elements in an array. These switching positions can also be treated as different patterns.

Extending the problem description above we now assume that not only one hole in a paper strip or one switch in an array is allowed, but $n$ holes or switches may be activated at once. Then the number of possibilities can be followed from the laws of combinatorial math

$$m = \frac{N!}{n!\,(N-n)!} = \binom{N}{n}. \tag{3.46}$$

This approach can be extended to a two-dimensional field of switches (Fig. 3.9). For $N$ fields and $n$ closed switches the amount of combinations can be derived from the equation above. If the number of closed switches varies between $n = 0$ and $n = N$, we obtain the number of possibilities by summation

$$m_s = \sum_{n=0}^{N} \binom{N}{n} = 2^N. \tag{3.47}$$

This example can be transferred to integrated circuits, where the switches correspond to transistors. The interconnect wiring is irrelevant for the calculation of the information content, because it is static. The sequential logic system of a computer changes its state over time. The various states of this machine can be interpreted as data or different patterns. Thus the sequence of events inside a computer system can also be treated with the term of information. In the following we will examine the transmission and processing of data in more detail.

Information theory models can also be found in other fields of physics, for instance in thermodynamics: If we consider an ideal gas and divide its volume into $N$ cells, where every cell can only accept one particle and we want to

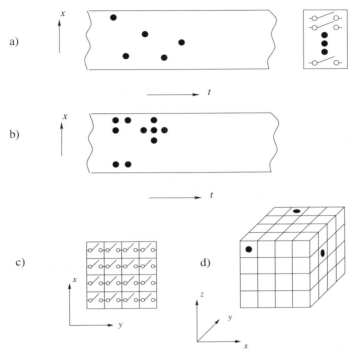

**Fig. 3.9.** Different patterns of data: (a) punched paper tape with single holes activating switches, (b) hole combinations as information, (c) two-dimensional switch array, (d) three-dimensional distribution of particles in a segmented volume

distribute $n$ particles in this volume, then the information pattern becomes a three-dimensional structure (Fig. 3.9). The number of possible combinations of distributing $n$ particles to $N$ cells is given by (3.45). Because $n$ and $N$ are very large we can reduce the $N!$ term by approximation. First we apply Stirling's equation

$$\ln N! \approx N \cdot \ln N - N + \frac{1}{2} \ln (2\pi N). \qquad (3.48)$$

Then we reduce this expression by neglecting the last part for $N \gg \ln N$ and by using $n \ll N$

$$\ln m \approx n + n \cdot \ln \frac{N}{n}, \qquad (3.49)$$

where $m$ is the number of different patterns that can be found if $n$ available particles are applied to an ideal gas of the volume $V$. This equation represents also a fundamental microelectronic relation.

For the thermodynamics Boltzmann demonstrated that the entropy $S = \frac{W}{T}$ of an ideal gas can also be defined by the number of microelectronic states,

$$S = k \cdot \ln m, \tag{3.50}$$

where $k = 1.38 \times 10^{-23}$ J K$^{-1}$ is the Boltzmann constant. The entropy of a closed system is constantly increasing, because disorder grows under the influence of heat. Based on this thermodynamic consideration the minimum energy that is needed to process 1 bit of information equals

$$W = 2kT \ln 2. \tag{3.51}$$

This relation derives from $S = \frac{W}{T}$, (3.45) and (3.50). It is noticeable that this relation is formally equal to (3.47). This prompted Shannon to give the name entropy also to the information content $H_0$. Shannon develops the information theory from mathematical abstractions, but it was Brillouin's idea to transfer his considerations into physics. The ideas were further developed by Weizsaecker and applied to microelectronics and computer science by Meindl and Keyes. We have to take great care in the choice of the entropies sign: Because the entropy decreases by receiving information, information must be compared to negative entropy. This is why information is also called negentropy.

### 3.2.2 Data Processing

In the second half of the last century information theory focused on information transmission. This process is illustrated in Fig. 3.10 and consists of a transmitter, a transmission channel that is usually disturbed, and a receiver. The processing blocks represented in this diagram play an important role in microelectronics. The transmitter and also the receiver have to encode, transform, and decode the information flow.

Transmitter and receiver also occur in information processing. In this case they are generally presented in one person, namely the computer user. The

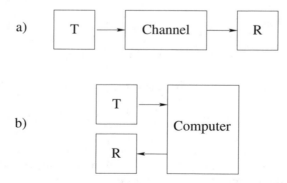

**Fig. 3.10.** Block diagram of (a) an information transmission system and (b) an information processing system. The channel, as well as the computer may be affected by noise

transmission channel is represented by the computer, which not only transmits information, but also transforms, i.e. changes it (Fig. 3.10).

An important characteristic of every transmission is the information quantity $M$ measured in bits

$$M = \frac{t_t}{t_s} n \cdot \operatorname{ld} m, \qquad (3.52)$$

where $t_s$ is the time needed for one transmission step. The total quantity of information can be represented as an information cuboid illustrated in Fig. 3.11. The following measures are plotted on the axes of this shape: The complexity $\operatorname{ld} m$ of the symbols, the number $n$ of symbols, and the number $\frac{t_t}{t_s}$ of steps. The information flow is defined as the amount of steps per unit of time

$$I = \frac{dM}{dt}, \qquad (3.53)$$

measured in $\frac{\text{bits}}{\text{s}}$ or baud.

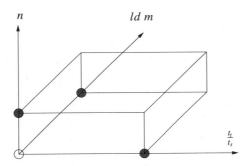

**Fig. 3.11.** Cuboid for the illustration of information quantity. The following measures are assigned to the three axes: degree of parallelism (bandwidth), transmission duration, and complexity of the patterns

The goal of a transmission is to forward this set as complete as possible. In contrast to a transmission, processing aims to selectively transform the information. The performance of information processing is therefore not measured in baud, but in MIPS (millions of instructions per second). In comparison to the transmission process a similar cuboid can be constructed. The axes for this case are: the number of pattern processed in the time $t_c$, their complexity, and the number of patterns per processing step.

$$M = \frac{t_c}{t_s} n \cdot \operatorname{ld} m. \qquad (3.54)$$

Because a computer system changes the processing data, the information sets at the input and the output may have a different size.

Most microprocessor systems known today are based on the so-called Turing machine: Data is read from and written to a memory, and a processor

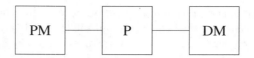

**Fig. 3.12.** The basic concept of information processing: A processor $P$ processes data from the memory $DM$ by a program stored in $PM$

processes the information (Fig. 3.12). This concept gains its outstanding importance not without the thesis of Church, which says: The Turing machine is universal computational, i.e. all problems that can be solved by some method of computation can be solved by this machine. For nanoelectronics this statement is fundamental, because it means that every thinkable method of storing information and virtually every set of operations on this information is sufficient to solve any computational problem.

In Chap. 1 we did see that some so-called hard problems could not be solved in this way due to a lack of time. This is why we have to apply concepts of parallel computation. Thereby we have to use new concepts that are often predetermined by specific technologies. A survey of different technological fields is given in Fig. 3.13. In the following chapters examples of these four elementary concepts will be presented.

When considering information processing we can choose between digital and analog technology implementations. Thereby we have to ensure that no information loss occurs. The sampling theorem gives an answer to this fundamental question (Fig. 3.14). It can be shown that the complete information content can be transmitted by discrete samples, if only the sampling rate is chosen adequately. Current, voltage, and power values are sampled in the case of electronics. The time difference $\Delta t$ between these samples has to be shorter than the halfperiod of the highest frequency $f$ or bandwidth $\Delta f_0$ a signal consists of.

$$\Delta t < \frac{1}{2\Delta f_0}. \qquad (3.55)$$

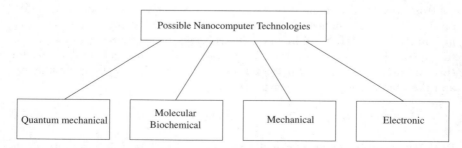

**Fig. 3.13.** Technological concepts for nanoelectronics and nanosystems

This relation can be made plausible by considering a Fourier transformation: The signal sine-wave oscillation with the highest frequency has still to be captured.

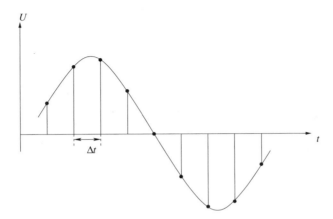

**Fig. 3.14.** Illustration of the sampling theorem. In order to capture the signal completely it is sufficient to scan the signal every $\Delta t$ that corresponds to a sine-wave oscillation with the highest frequency

Another problem is to determine the accuracy of an analog value represented in digital form. But an analog value can also be represented only with a restricted degree of accuracy. If a digital signal of $m$ discretization steps is 'by means of measuring technology' indistinguishable from the analog signal, the number of discretization steps is high enough to represent the signal without information loss, and both approaches are equivalent solutions.

The signals we can observe in nature are, as a rule, analog. For the transition to digital representation we have to apply an analog to digital converter (ADC). Nowadays integrated circuits development trend toward digital implementations, because digital switching elements are easier to construct and put together. In many cases the transmission quality can be significantly increased by using digital technology, because a high degree of precision can be reached by utilizing long data words. Nevertheless, every system on a chip generally needs an analog part, because at least an input circuitry has to perform some kind of analog to digital conversion.

## 3.3 Summary

Classic models of electrical engineering are generally sufficient for microelectronic applications. By introducing nanoelectronics, it is necessary to use quantum-mechanical concepts. Different physical values are quantized, which leads to novel effects. Basic measures of information theory also exhibit some kind of quantization. These effects not only offer new ways of technological developments, they also erect limits to nanoelectronics.

# 4
# Biology-Inspired Concepts

The brain of any living being can be considered as a high-performance information-processing system. Besides their facilities, the underlying architectures are of general interest for the development of nanoelectronics and they might be helpful to set a goal on nanoelectronics.

Biological concepts can also be copied. In the following, three examples will be discussed that have been selected from many cases. The first example deals with the co-integration of biological neurons and MOS circuitries on the same substrate. The second example comprises the imitation of biological neurons via VLSI circuitries. An artificial neuronal network with local adaptation and spatial interconnected data streams is revealed by the third example.

## 4.1 Biological Networks

### 4.1.1 Biological Neurons

Neurons consist of a nucleus and long dendrites reaching to neighboring cells. The signals travel via the dendrites to other neurons (Fig. 4.1a).

The axon forms the output of the neuronal cell. As a rule, the axon terminates in a synapse that establishes the connection to other neurons. In this context, the so-called dendrites are connected to the synapses. They are in charge of the input signals of the neuron. A more precise description of the biological neuron's setup is available in the relevant literature [19].

Detailed observations of biology confirm the spread and storage of information in neuronal networks. The information spread takes place along the axons, whereas information is stored within the neurons. Short-term storage is realized in terms of electrical signals, while long-term storage is based on biochemical modifications in the synapses.

Within the synapse an axon meets a dendrite; the junction comprises a presynaptic membrane, the synapse gap and the postsynaptic membrane (Fig. 4.1b). Roughly 1 000 vesicles are located at the end of each axon. They

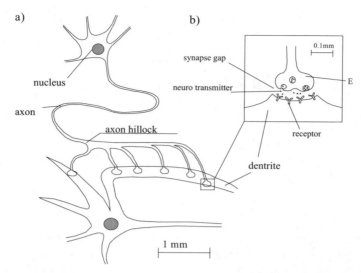

**Fig. 4.1.** Schematic view of a biological neuron (a) and square section of a synapse (b)

comprise about 10 000 molecules of a transmitter substance and are of fundamental relevance for information storage. An incoming electrical signal on the axon pours the neurotransmitter molecules out of the vesicles into the synapse gap. Thus, the synapse is a simple biomolecular processor. According to Penrose et al., the action of a synapse lies in the range of a few quantums of action h. The energy that is transformed in relation to the information processing is very small and lies in the order of 10 fJ. Receptors of the postsynaptic membrane absorb the neurotransmitters and provoke a rising or falling voltage level in their accompanying neuronal cell. The coupling might be excitatoric or inhibitoric. Beyond a distinct threshold voltage level the neuron emits a signal via the axon [20].

Both learning and drugs modify the synaptic connections of a neuronal cell. According to a generally accepted theory, adaptation takes place in a form that the adapted events run automatically (Hebbian rule). Thus, these events are considered to be dominated like writing or swimming. The adaptation process accounts for the effort of learning. On the contrary, drugs offer fantastic worlds and such imaginations appear without any effort, since the synapses are stimulated to a great extent without any control. This causes fantasies within the brain that make people become addicted.

The dimensions of neurons are in the order of millimeters, however, axons can measure several centimeters and even in some cases more than one meter. Biological neurons have relatively large dimensions in comparison to integrated CMOS circuits. However, the three-dimensional structure of neu-

4.1 Biological Networks    63

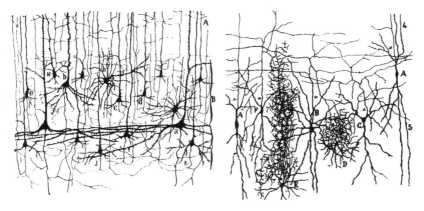

**Fig. 4.2.** Photograph of biological neurons that have been preserved by Caja [19]. The samples show that there are quite various neurons

ronal networks is very advantageous, in contrast to the presently limited two-dimensional integrated circuits. This limitation not only accounts for a low packing density, but also limits the wiring and architectural facilities.

Figure 4.2 illustrates preserved neuronal cells of the cerebral cortex. According to the applied preparation method of Caja, only 1% of neuronal cells appear in the clear arranged photograph. Thus, the complex interconnections and the high packing density can only be estimated. Furthermore, the photograph reveals the existence of different biological neurons with individual complex interconnects. Up to now artificial neuronal networks do not account for these differences.

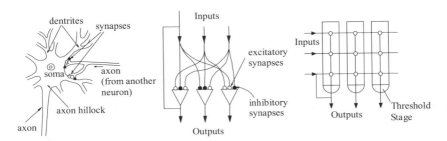

**Fig. 4.3.** From the biological neuron via the threshold gate to the matrix architecture

Simple architectures for artificial neuronal networks derive from biology-based neuronal networks (Fig. 4.3). As described before, neurons can be treated in a first-order approximation as threshold gates. Two adjacent gate levels are fully interconnected. This architecture is equivalent to a matrix configuration. Each column corresponds to a neuron and each dot represents the synaptic coupling. In terms of artificial neurons, these couplings are refered to as weights, which evaluate the input signals.

Additionally, distinct biology-based networks comprise feedback loops between different gate levels. In this context, research has been focusing for the last twenty years on artificial neuronal networks that originate from this structure. However, one has to keep in mind that similar results can be gathered from physics in terms of self-organization problems, e.g. the Ising model. These architectures are of fundamental relevance for nanoelectronics, since data processing takes place in a parallel and local fashion. The borders between memory and processor are vanishing.

### 4.1.2 The Function of a Neuronal Cell

The dynamic behavior of a single biology-based neuron is very interesting: Investigations of the axon's response time surprisingly show that the relatively high output signal gets activated if the input signal crosses a distinct threshold value. The output signal itself does not depend on the slope of the input signal (Fig. 4.4). Within the experiment the internal potential of the axon gets fixed so that the current through the membrane becomes measurable [19]. The question arises, how can this behavior be explained?

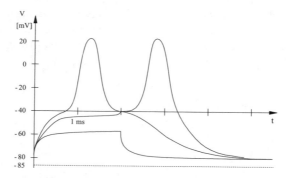

**Fig. 4.4.** Impulse response of the axon. The axon only gets activated beyond a distinct threshold value

First, we will derive the resting potential of the neuron membrane of about - 85 mV (Fig. 4.5). The cell membrane of a neuron exists as a double molecular layer. Its thickness measures roughly 5 nm, which results in an electric field of about $1.7 \times 10^5$ $V/cm$. This value is about a factor of ten below the maximum electrical field of a CMOS transistor. The membrane is located in an aqueous solution of the body with relative permittivity of $\epsilon_w = 81$, whereas the inside of the membrane has a relative permittivity of $\epsilon_i = 2$. Thus, the membrane comprises a potential barrier for charged molecules, like Na ions, of about $2.3\,eV = 100\,kT$.

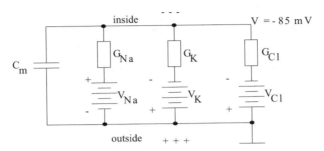

**Fig. 4.5.** Resting potential of the neuron membrane: Sodium ions and potassium ions are transported by an ion pump. Due to the density ratio, the neuron membrane shows a potential difference

During the metabolism of living, neurons ions are pumped through the membrane. Thus, different ion densities on the inside and outside of the neuron can be traced: For instance, the density ratio of K ions between the inside and outside is 400 and 10 mM/l, whereas the density ratio of Na ions amounts to 50 and 460 mM/l. Equivalent to a doped semiconductor the different densities result in a diffusion voltage:

$$V_D = \frac{kT}{q} ln \frac{N_{in}}{N_{ex}} \qquad (4.1)$$

$N_{ex}$ is the density of the extracellular aqueous solution, whereas $N_{in}$ accounts on the density in the cytoplasm. Thus, the potential difference of the K ions amounts to $V_K = -92\,mV$ and the potential difference of the Na ions is $V_{Na} = +55\,mV$. Furthermore, Cl ions cause an additional potential difference.

The capacitance $C_m$ of the equivalent circuit diagram in Fig. 4.5 represents the membrane that is connected via internal resistances to the diffusion voltage supplies. (Further iontypes are also involved, but are neglected here.) With respect to the internal resistances, the resting potential amounts to $-85\,mV$. Ingestion delivers the energy to maintain this state.

Next, the dependence of the internal resistances on the potential will be analyzed. Surprisingly, the conductance varies exponentially with the potential for low potential values, whereas the conductance saturates for high potential values (Fig. 4.6). The conductance depends on the number of active ion channels. The number of active channels varies with respect to the iontype and the potential (Fig. 4.6).

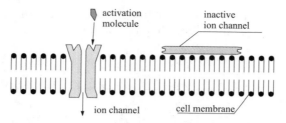

**Fig. 4.6.** Cell membrane, approximately 5 nm thick, with an active and inactive ion channel

The degree of alternation based on the Boltzmann law: $W_t$ describes the transitional energy and $N_O/N_C$ represents the ratio of opened and closed channels:

$$\frac{N_O}{N_C} = e^{-\frac{W_t}{kT}}. \qquad (4.2)$$

The transitional energy depends on the number n of electrically charged particles that pass through the channel:

$$W_t = W_O - qnV_m. \qquad (4.3)$$

In this context, the normalized conductance is of fundamental relevance and based on the ratio between the number of opened channels $N_O$ and the overall number of channels $N$:

$$\frac{\frac{N_O}{N}}{1-\frac{N_O}{N}} = e^{-\frac{W_O}{kT}} e^{\frac{qnV}{kT}}. \qquad (4.4)$$

Equation (4.4) describes the exponential dependence of the experimental data in Fig. 4.7. The ion channels control the current over the range of several orders of magnitudes, which is equivalent to electronic switching devices. The slope and the location of the conductance-voltage characteristics is of fundamental relevance for the correct operation of the neuronal cells.

From the already derived basics we might answer the next question: Why does the neuron behave like a threshold gate (Fig. 4.4)? The contact potential of the quiescent state amounts to $-85\ mV$. Electrical pulses below $-40\ mV$ cause the typical charging and discharging of the capacitance $C_m$. An external excitation beyond $-40\ mV$ results in an output pulse of $+20\ mV$, which is independent of the intensity of the excitation. The exceeding of the threshold value initiates a complex program in the cell membrane: First, a rapid $Na^+$ current drives the potential towards $+20\ mV$. A delayed and slower $K^+$ current forces the potential back to its initial state.

This model also explains the signal propagation along the axon (Fig. 4.8). Axon potentials that exceed $-40\ mV$ cause ion channels. Thus, Na ions move

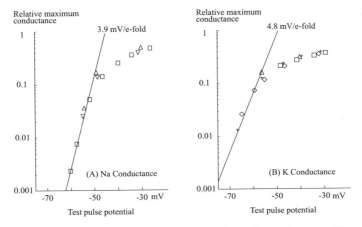

**Fig. 4.7.** Conductance-voltage characteristics of a cell membrane with respect to the conductance of Na ions and K ions

towards the axon's inside and change the potential of the membrane. With a slide delay further ion channels develop and K-ions put the potential back to its initial state. It takes some time until the resting state is reached again. During this period the signal propagates along the axon. Due to the fact that the axon has to pass through this regeneration period the propagation direction is unambiguous.

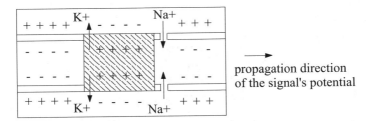

**Fig. 4.8.** Schematic view of an axon and the propagation direction of its signal

## 4.2 Biology-Inspired Concepts

Consistently, engineers are inspired by biology. Such approaches like jet propulsion are appropriate, however, the helicopter's propeller is very efficient but does not appear in nature. The following three examples demonstrate the variety of possibilities.

### 4.2.1 Biological Neuronal Cells on Silicon

Initially, an engineer would probably consider the growth of neuronal cells on silicon and their combination with integrated circuits as an absurd and impossible dream. Nevertheless, experiments of P. Fromherz (MPI) show that neuronal cells from the brain continue to grow in a nutrient solution. Normally, this growth process takes place in a glass bowl under uncontrolled circumstances. However, if the growth process takes place on top of a structured silicon substrate (e.g. etched channel structures in a silicon oxide layer) the nerve fibers grow along these channels. Channel branches make the nerve fibers split and the end of a channel stops the growth process (Fig. 4.9). According to this method, the shape of the neuronal network is determined by the silicon surface structure [21].

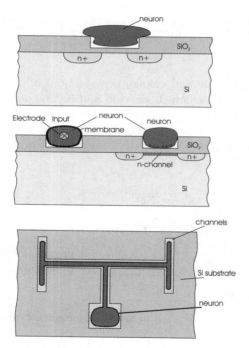

**Fig. 4.9.** Biological cells grow on top of a silicon substrate. The electric fields of the neurons interact with MOS structures

Electrical pulses can propagate along the nerve fibers. Such pulses might originate from a synapse or might come from an external sensor. Other forms of excitations are: Direct signal feeding and influenced charge in the nerve fiber. Thus, the nerve fiber is interconnected with an electric circuit. The electric pulse follows the nerve fibers that are located in the silicon channels.

MOS structures are capable of picking up the signals of the biological network, which enables the interconnection with integrated circuits. For instance, the charge of a nerve fiber can influence a charge at the surface of the silicon, which creates a conducting channel between the source and drain area of a MOS structure (Fig. 4.9).

Up to now, the packing density of this technology is very low and such concepts are not suitable for computer applications. They might become interesting for medical applications, since the whole human body is accessible via its neuronal network. Status information of the different organs might be gathered by tapping the neuronal network, which could save the money spent huge medical equipments. Today this is still a vision, but in the future it might help curb the costs of the public health system. Additionally, such concepts are interesting for artifical limbs. However, they are also interesting for nanoelectronics because this hybrid technology combines biological neuronal networks with electrical circuits. The combination is advantageous, since the complex behavior of the biological neuronal cell becomes integrated in a technical system.

### 4.2.2 Modelling of Neuronal Cells by VLSI Circuits

Nanoelectronics tries to copy biological concepts for information-processing systems. The following example reveals the reproduction of a neuronal cell with its exponential input-output characteristics on the basis of MOS transistors [22].

The above-described behavior of the neuronal cell is equivalent to the MOS circuit in Fig. 4.10. The input current $I_E$ charges the capacitance $C_1$. An adequate potential change lets the subsequent latch switch. The output voltage is almost independent of the input voltage, however, it depends on the supply voltage that corresponds to the membrane potential. Simultaneously, transistor $T_2$ is activated and discharges the capacitance $C_1$. After a time, the potential has decreased sufficiently and the latch switches back to its initial state and the procedure can start over again. It follows from the equivalent circuit diagram: The more intensive the input signal, the faster the capacitance $C_1$ gets charged and the more dense the pulse train at the output. From the viewpoint of microelectronics this part of the axon represents the functional integration of a multivibrator. In terms of biology, the switching devices consist of a cell membrane and an aqueous solution with Na and K ions.

If the MOS transistors are operated in the subthreshold domain, not only the function but also the exact characteristics of the membrane can be imitated, since the transistors show in this domain (just like the biological membrane) an exponential characteristic. The simple MOS transistor model assumes beyond the threshold voltage a square current-voltage dependence. However, below the threshold voltage a subthreshold current is observable.

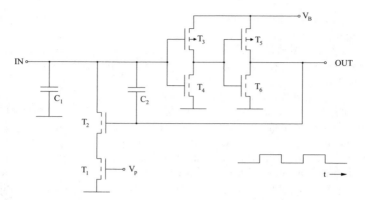

**Fig. 4.10.** Equivalent circuit diagram of a neuronal cell and the output signal of the threshold gate behavior

Figure 4.11 illustrates the cross section of a p-MOS transistor and will be helpful to explain this current.

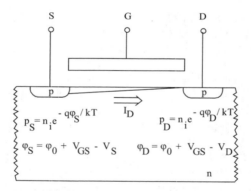

**Fig. 4.11.** Cross section of a p-channel MOS transistor

Similar to the pn-junction, the charge densities of the p-MOS transistor channel depend on the surface potentials $\phi_S$ and $\phi_D$ of the source and drain areas:

$$p_S = n_i e^{-\frac{q\phi_S}{kT}}, \quad p_D = n_i e^{-\frac{q\phi_D}{kT}}. \tag{4.5}$$

$n_i$ represents the charge density at the Fermi potential. The surface potentials follow from the applied source, drain, and gate voltages:

$$\phi_S = \phi_0 + (V_{GS} - V_S), \quad \phi_D = \phi_0 + (V_{GS} - V_D). \tag{4.6}$$

$\phi_0$ is the built-in potential of the gate insulator that governs the threshold voltage $V_{Tp}$. The factor m accounts for the fact that normally the applied voltage does not completely reach the semiconductor surface. To simplify matters, this factor will be chosen as one in the following. The channel current is based on diffusion and the charge gradient can be approximated form (4.5) and the channel length L:

$$\frac{\delta p}{\delta x} = \frac{p_D - p_S}{L}. \quad (4.7)$$

The diffusion current is proportional to the charge gradient and amounts to:

$$I = -qWD\frac{\delta p}{\delta x} = qD_P\frac{W}{L}(p_S - p_D). \quad (4.8)$$

It follows from the above:

$$I = I_0 e^{-\frac{V_{GS}}{V_T}} \left( e^{-\frac{V_S}{V_T}} - e^{-\frac{V_D}{V_T}} \right). \quad (4.9)$$

In this context, $I_0$ comprises all stationary values. Choosing the source contact as reference potential ($V_S = 0$), the equation simplifies to:

$$I = I_0 e^{-\frac{V_{GS}}{V_T}} \left( 1 - e^{-\frac{V_{DS}}{V_T}} \right). \quad (4.10)$$

Equation. 4.10 describes the subthreshold voltage dependence of a MOS transistor. Its current changes by orders of magnitude, due to the exponential current-voltage characteristics (Fig. 4.12).

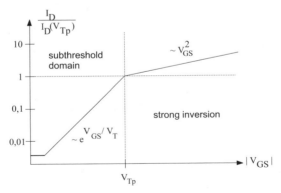

**Fig. 4.12.** Schematic view of the subthreshold current-voltage characteristics (left side) and the strong inversion characteristics (right side) of a p-channel MOS transistor

In terms of small current values, the subthreshold current is dominated by leakage currents, whereas high current values are limited by the resistance of the inversion layer (classical model).

On the basis of the subthreshold current model, different complex VLSI systems like the electronic retina and the electronic cochlea have been already developed. C. Mead has been primarily involved in this area and has implemented biology-inspired concepts into analog MOS circuits [19]. Clearly, these strategies are also of fundamental relevance for nanoelectronics.

### 4.2.3 Neuronal Networks with local Adaptation and Distributed Data Processing

An important objective of nanoelectronics is a local data processing structure in order to avoid long wiring distances within the chip. This constraint might be met with biology-inspired neuronal networks. In contrast to many artificial neuronal networks, the rule of Hebb is a simple model for the description of neuronal cells and is based on local adaptation. The adaptation of an advantageous network would only be based on input and output signals (Fig. 4.13). Nevertheless, such a network can be arranged in a matrix and modular way.

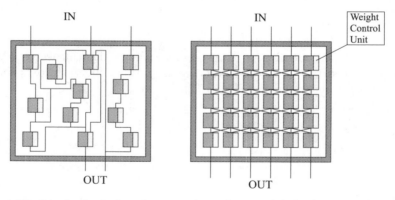

**Fig. 4.13.** Local adaptation does not depend on a global adaptation logic, but has distributed weight Control units. Within such a network, the neurons can be arranged in a regular manner

The essential difference in comparison with conventional neurons (Fig. 4.14) is the local adaptation process that is based on a weight control unit. According to the rule of Hebb, adaptation only takes place if the input and output signals coincide. Otherwise the information falls into oblivion, which means the weights are weakened.

Figure 4.15 illustrates the timing of the adaptation process: First, a weight is assigned to the input signal. During the following steps, the weight of the

4.2 Biology-Inspired Concepts    73

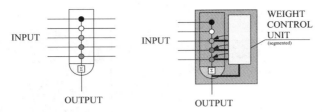

**Fig. 4.14.** Schematic view of a conventional neuron as well as the schematic view of a neuron that is capable of local adaptation.

Hebb synapse develops on the basis of the coincidence of the input and output signal. Further adaptation steps might increase the initially small weight. This results in a complete conditioning, namely the learning target, like cycling. Usually, the Hebb neuron is not based on a narrow timing and filters have to be applied to enlarge the time signals. Thus, the probability for a temporal coincidence is higher.

The schematic diagram in Fig. 4.16 reveals the function of a single neuron and comprises the above-described local adaptation options. This neuron type is also based on synapses with inputs that can have an excitatory or inhibitory impact. A fundamental feature of this neuron type is the veto-input that blocks the synapse input. Therefore, a command with higher priority might disable the neuron, e.g. for safety reasons. The so-called veto-synapse additionally limits the interconnection region of the neurons.

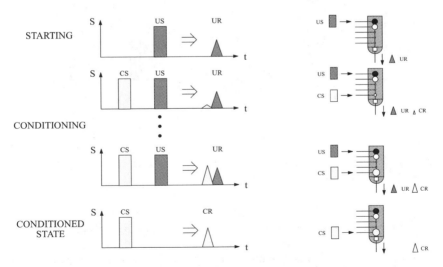

**Fig. 4.15.** Hebbian learning: Timing of the local adaptation process

74    4 Biology-Inspired Concepts

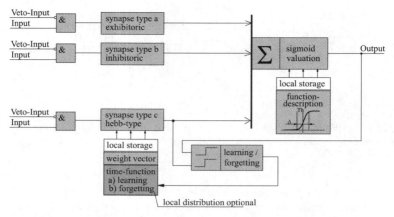

**Fig. 4.16.** Schematic diagram of a single neuron with local adaptation options

In general, a few neurons with only some inputs can handle quite complex procedures. For instance, only seven neurons already control a wheelchair. For comparison, a fly needs only 17 neurons for its motor activity.

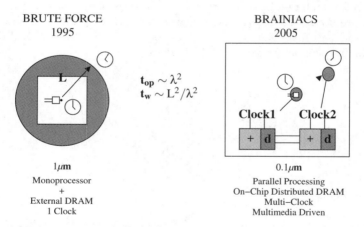

**Fig. 4.17.** Increasing packing densities require new architectures that might be inspired by neuronal networks

Nanoelectronics will comprise millions of neurons. Nevertheless, it will be problematic for such a huge system to automatically adapt an efficient architecture. But certainly, this amount of neurons will be capable of very complex procedures like the human brain (Fig. 4.17). Their efficiency will always be higher than conventional networks, if they have to deal with indefinable or transient environments. These concepts are also interesting since data is not stored in a single dot as described in Fig. 3.12, but dynamically interconnected in three-dimensional space. The structuring process might be based on self organizing principles. In this context, research is still at its very early stage. Certainly, the results will be very interesting for nanoelectronics.

## 4.3 Summary

For orientation purposes, it is very interesting to take a closer look at biological neuronal networks, since they show a high system performance. Therefore, biology-inspired electronics is an interesting alternative to classical approaches, but has serious disadvantages like the lack of robustness and low reliability, etc. In the long term, these disadvantages will make it difficult to transform this area of research into any application. However, it will have a fundamental impact on nanoelectronics.

# 5

# Biochemical and Quantum-mechanical Computers

Information processing requires new concepts for the solution of complex problems that will become even more complex in the future. All solutions for these problems envisioned today use parallel computation. However, the traditional microelectronics does not offer appropriate architectures nor efficient wiring technologies for parallel processing. Therefore, these problems are a challenge for nanoelectronics. The difficulty of finding appropriate solutions for these problems gives ample scope for new, unusual ideas, too.

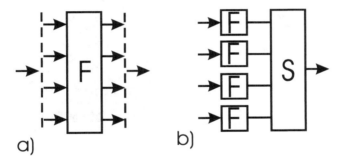

**Fig. 5.1.** Two unusual concepts for parallel processing: (a)quantum-mechanical computer, (b) DNA computer, processing unit (F), selector (S)

Unusual concepts are biochemical computers, such as the DNA computer, and the quantum-mechanical computer (Fig. 5.1). The following presents both of these concepts although their realizations are still far away. Both examples show that it is important to enlarge the scope beyond the nanoelectronics implemented in solid-state materials. In this challenging case we have to consider all possible ways that lead to an efficient parallel processing [23].

## 5.1 DNA Computer

### 5.1.1 Information Processing with Chemical Reactions

If we regard the excellent solutions for information processing of nature it is obvious to copy them. An example of this is the DNA computer that carries out the computations by chemical reactions [24]. The data are molecules inside a test tube. By adding extra substances the intended reactions are carried out whereas a robot can automatically perform all mechanical operations and measuring techniques (Fig. 5.2). In this concept the single molecules are the nanocomputers producing a huge number of solutions. The problem is to find the correct solution among the enormous number of results.

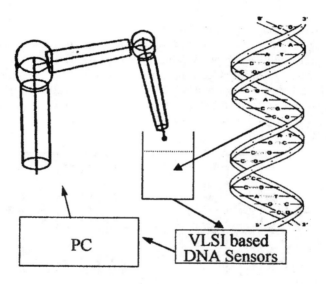

**Fig. 5.2.** Principle of a DNA computer: The information process occurs in a test tube, a robot can handle the operations automatically

Adleman [25] proposed the DNA computer and tested it in the lab. He used DNA molecules (deoxyribonucleic acid) - the basis of life - which is also known as double helix in the cells of living beings. DNA is a double-stranded molecule made by stringing four nucleotides together. The four nucleotide building blocks are A (adenine), T (thymine), G (guanine), and C (cytosine). By using this four-letter alphabet, DNA can store genetic information that is manipulated by living organisms. The mode of operation is close to that of silicon-based computers with binary encoding, but DNA stores the information by using four symbols instead of two.

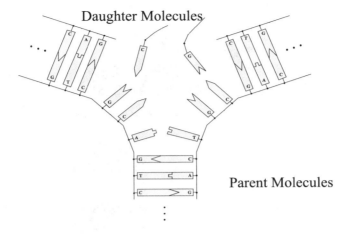

**Fig. 5.3.** Part of a DNA molecule, just in the replication process

A DNA molecule consists of two strings where each base on one string has a complementary base on the other string: A matches with T, and C matches with G, and vice versa. During the replication process only the corresponding molecules fit together, as Fig. 5.3 shows. The density that can be reached with DNA molecules is impressively shown by biology: The density of $10^{21}\ bit/cm^3$ or $1\ bit/nm^3$ is much higher than those of a neural net in the brain. As a comparison the information of $10^{21}\ bit$ corresponds to the information of about one trillion CDs.

### 5.1.2 Nanomachines

The DNA polymerase, which is an enzyme, can produce a second DNA strand that is complementary to an existent one. In this way the DNA strand gets a partner - we call this knitted - and the original molecule has doubled. The polymerase is a nanomachine consisting of only one molecule. It sticks on a given DNA strand, slips along the strand, reads one after the other base and inserts the complementary base including the part of the backbone to the new DNA strand. It operates as an inverter and writes a complementary copy of the strand (Fig. 5.3).

That is the story from the biological point of view, the connection to the computer world is done via the Turing machine (Fig. 3.12). In a simple form a Turing machine consists of two storage tapes and a processing unit that moves along the tapes. One tape acts as an input tape, which is only readable, whereas the processor unit can read and write to the output tape. According to this model the DNA polymerase operates in a similar manner to the Turing machine.

At a first glance the idea of an enzyme, here the polymerase, acting as a processing unit is obvious. However, it is not as simple as that - some processes

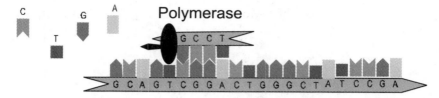

**Fig. 5.4.** Polymerase as a copying machine, it works as an inverter and produces a complementary strand to the original one

have to be added to get the desired solution. The following indicates the most essentials steps:

1. Combination of complementary strands: If a DNA strand meets its complement they are twisted to a double helix.
2. The polymerase as a copying machine. The existence of a so-called primer is necessary for the correct operation of the copying machine. The primer is the starting point for the knitting. The polymerase extends each appearing primer in a preferential direction to a complementary strand.
3. Ligases are enzymes that combine two DNA strands to a longer one.
4. Nucleases cut the DNA strands at defined positions.
5. The gel electrophoresis sorts the DNA molecules with regard to their lengths.
6. The DNA synthesis produces DNA molecules with defined sequences and strand lengths. A test tube with $10^{18}$ molecules with a length of about 50 base pairs cost 40\$ in 2000.

Adleman solved a traveling salesman problem (TSP) with the help of a DNA computer described above. His computer does not consist of any silicon, instead of silicon it uses some rows of test tubes, partly empty while others are filled with DNA strands. The following 5-step algorithm was proposed by Adleman for solving the problem at the molecular level:

1. Generate random paths through the graph.
2. Keep only those paths that begin with the start point and end with the end point.
3. If the graph has $n$ nodes, keep only those paths that include all $n$ nodes.
4. Keep those paths that include the start- and end-point and pass all nodes only once.
   At least one solution is available.

This algorithm has to be implemented with DNA according to the following steps:

1. Synthesis of the desired strand.
2. Separation of strands according to their length.
3. Mix two test tubes into one to perform a union.

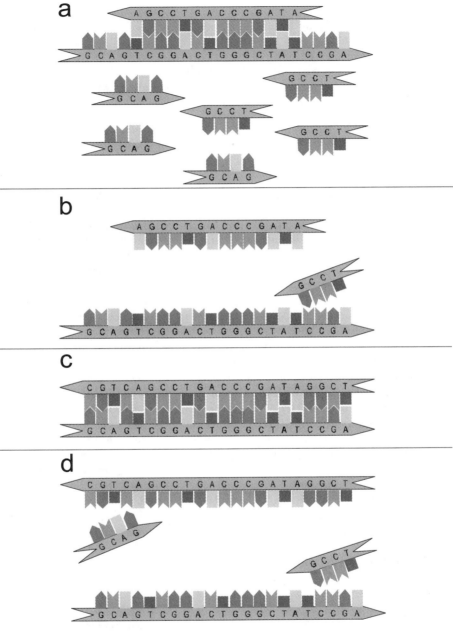

**Fig. 5.5.** Steps for duplicating and assembling the DNA strands

4. Extraction of those strands containing a given pattern.
5. Melting/annealing by breaking/bonding two DNA strands with complementary sequences.
6. Amplification by using polymerase to copy existing DNA strands.
7. Cutting of DNA strands with restriction enzymes.
8. Attach DNA strands with complementary endings with ligases.
9. Detection by confirming presence/absence of DNA in a given test tube.

Many combinations will be formed in test tubes. If billions of molecules and reactions are performed it is quite certain that the correct path will be generated and will be in the final mixture.

### 5.1.3 Parallel Processing

The last process steps have to find the correct solution among $10^{11}$ results by sorting the DNA strands according to the defined criteria. Actually the reaction was successful: After about 5 days the DNA computer came up with the correct solution. A high-performance computer needs much more time to solve this problem, roughly millions of years.

The selective reproduction is an important process step for an efficient computation on the DNA level as shown in Fig. 5.5. This step drastically increases the number of nanomachines operating in the proximity of the solution. The combination of DNA sequences, containing the way numbers and the complementary place-names, produces among others double-stranded molecules containing the correct start and destination address. Figure 5.5 depicts only the double strand for the Hamiltionan path that is the solution. The arrow at the backbone denotes the read direction. In order to copy the correct double strand, primers are added to the solution. These primers contain the name of the starting point (GCAG) or the complement name of the end-point (GGCT). In the first reproduction step the double strand is separated into two single strands. An appropriate primer attaches only to the longer strand (Fig. 5.5b). The polymerase produces a complete double strand from the single strand (Fig. 5.5c). If this new double strand is separated in repeated steps, further primers can attach at both strands (Fig. 5.5d) so that the number of these DNA strands doubles each step and exponential growth is obtained. This procedure only works if the strands start and end with the correct place names. This method drastically increases the number of solutions that can contain the correct solution. In addition, it also increases the probability of getting the correct solution.

A biomolecular computer on the basis of DNA accomplishes a massive parallel information processing. One small drop of a solution combines $10^{14}$ DNA strands per second. These processing steps consume only a small amount of energy, since the transportation of information consumes no energy (Chap. 11). $10^{19}$ operations consume 1 J. Present-day supercomputers consume 1 J for about $10^9$ operations. The laws of thermodynamics set a maximum limit on

$30 \times 10^{19}$ operations, so the DNA computer operates quite near at the physical limit.

At the present time biocomputers based on similar concepts are under discussion. They may store the information in proteins and will use biology-inspired architectures. Such visions may become realistic in the future. Again the TSP is a good example of the use of those concepts.

## 5.2 Quantum Computer

### 5.2.1 Bit and Qubit

In the recent past the quantum computer indicates new potentials for efficient parallel processing [26]. The following describes a vision of this concept especially with regard to nanoelectronics. This vision shows that a new concept can overcome the barriers of classical information processing. Quantum computing is also subject to the fundamental assumption that information is associated with physical quantities. Whereas the classical laws of physics precisely define the information processing, quantum computing uses quantum objects as the information carrier. In our opinion these objects are subject to uncertainty.

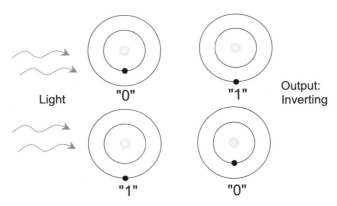

**Fig. 5.6.** Interaction of light waves with atoms, the structure inverts the input signal. However, this process describes no Qubit and is not based on the principle of quantum computing

At first glance we think that quantum computing applies quanta as bits. Figure 5.6 shows an inverter that uses an atom as a two-state system. If photons pass the atom it changes its state, i.e. the electron moves to adjacent energy states by absorption or emission of a photon. Although this structure works on the quantum level its mode of operation is not typical for quantum computers. The above-mentioned is only an example for the classical way of information processing at the atomic level and an example of the classical nanoelectronics. An inverter as such may be useful for input/output stages of a quantum computer. A similar structure on the quantum level offers the quantum cellular arrays, QCA, which chapter 10 will present.

In this context the so-called *EPR* experiment from Einstein, Podolski, and Rosen, which was the idea of an experiment, is worth mentioning and was recently confirmed. This experiment describes a two-electron system that is composed of two electrons with opposite spin. In the following we separate them and force them to fly in opposite directions. If we observe one of the electrons far away from the other and define the direction of its spin by observation of this electron, simultaneously the spin of the other electron changes to the opposite spin. This experimental result is inconsistent with our experience since the information about the spin must be transmitted between the two far-away electrons without any delay that would be faster than the speed of light. This effect is relevant for quantum computers and especially for quantum cryptography.

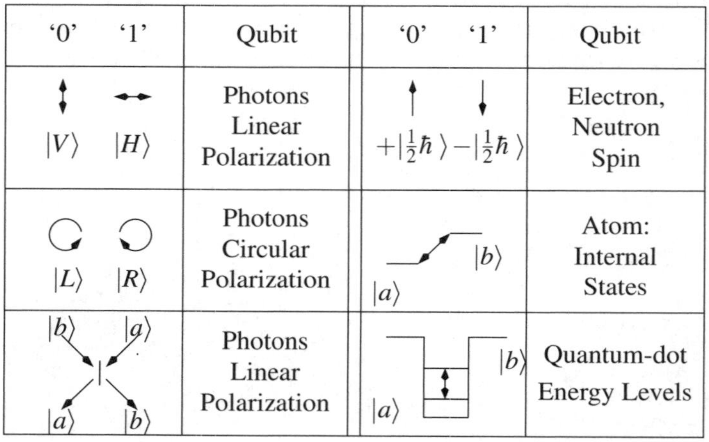

| '0' | '1' | Qubit | '0' | '1' | Qubit |
|---|---|---|---|---|---|
| $\|V\rangle$ | $\|H\rangle$ | Photons Linear Polarization | $+\|\frac{1}{2}\hbar\rangle$ | $-\|\frac{1}{2}\hbar\rangle$ | Electron, Neutron Spin |
| $\|L\rangle$ | $\|R\rangle$ | Photons Circular Polarization | $\|a\rangle$ | $\|b\rangle$ | Atom: Internal States |
| $\|b\rangle \ \|a\rangle$ $\|a\rangle \ \|b\rangle$ | | Photons Linear Polarization | $\|a\rangle$ | $\|b\rangle$ | Quantum-dot Energy Levels |

**Fig. 5.7.** Various realizations of a Qubit, H. Weinfurter and A. Zeilinger

The association between a bit and a two-state system is crucial for quantum computing. Examples of a two-state system can be different polarization states of a photon, the spin direction of electrons, the energy level of an atom or a quantum dot, or the modes of a beam splitter summarized in Fig. 5.7. If we denote both states of the system with $|a\rangle$ and $|b\rangle$ we can define the following identification:

$$0 \Leftrightarrow |a\rangle \quad , \quad 1 \Leftrightarrow |b\rangle. \tag{5.1}$$

This symbolic notation is common in physics and denotes the assignment of a logical 0 and 1 to physical quantities. The fundamental new concept of quantum mechanics compared to the classical laws of physics and data-processing technology is the coherent superposition of states. Therefore, the superposition of two states leads to the following result:

$$|\Psi\rangle = \alpha|a\rangle + \beta|b\rangle = \alpha|0\rangle + \beta|1\rangle. \quad (5.2)$$

From the viewpoint of physics this formula denotes that a bit is in the state $|a\rangle$ with the probability $\alpha^2$, corresponding to a logic 0, and additionally it is in the state $|b\rangle$ with the probability $\beta^2$, corresponding to a logic 1. A graphical illustration (Fig. 5.8) shows where the vector $\Psi$ is projected on the $|0\rangle$ and $|1\rangle$ axis. The value of the bit is undetermined before observation, which yields one of the two results with the given probabilities. Physicists call this effect coherence.

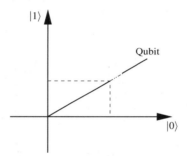

**Fig. 5.8.** Representation of a Qubit as a vector

However, if the described effect is only a mixture of various states no additional information content emerges. In our case this effect is a coherent superposition representing not only the states 0 and 1 but also intermediate states. Using this we get new operations of data processing that extend the possibilities of classical physics. A bit represented by quantum-mechanical laws is called a *Qubit*. It is described by a vector in the so-called Hilbert space.

### 5.2.2 Coherence and Entanglement

The coherent superposition of several Qubits is much more interesting than a single Qubit. As an example a three-Qubit string is represented by a three two-state systems

$$|010\rangle = |a\rangle|b\rangle|c\rangle. \quad (5.3)$$

The important point is the correlation of the string to a unique state (one vector) in the Hilbert space. In a simple case the three Qubits may be in the following state

$$|000\rangle + |111\rangle/\sqrt{2}. \tag{5.4}$$

Equation (5.4) assigns all three Qubits either the value 0 or the value 1 with a probability of 0.5. This result is fundamentally new in contrast to the classical physics where the bits are independent of each other. It is also fundamental that the value of one of the three Qubits is determined from the beginning. The observation of one of the three Qubits stops the uncertainty immediately and forces the other Qubits to specific values, which follow from (5.4). This characteristic of quantum mechanics is called *entanglement*. The advantage of this effect is that the measurable state of all three Qubits appears simultaneously and not due to complex coupling between gates.

The following numerical example demonstrates the potential of quantum computing. In the classical way we can store about $10^{80}$ bits if all $10^{80}$ atoms of the universe are used. By using Qubits we only need 266 atoms to store the same amount of information.

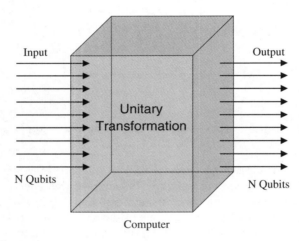

**Fig. 5.9.** Principle of a quantum computer, C. Bennett, IBM

### 5.2.3 Quantum Parallelism

A quantum computer has a number of Qubits as inputs and as outputs. The outputs are derived from the inputs on the basis of quantum-mechanical operations. Figure 5.9 shows the principle of such a simple quantum computer. In the simplest case the input state is the product of $N$ Qubits. The unitary transformation produces a well-defined product state at the output. With help of the superposition any input state is the coherent superposition of many product states, where each state can have the probability of $1/N$:

$$|Input\rangle = \{|Input_1\rangle + |Input_2\rangle + \ldots\}/\sqrt{N}. \tag{5.5}$$

The entanglement produces a superposition of all states that correspond to the respective input states. A quantum computer carries out a large number of computations simultaneously for a large number of various input states. This is called massive quantum parallelism.

Real-world data processing also requires nonlinear elements for logical interconnections. In this case the coherent modification of one Qubit is due to the state of another Qubit. For example, a series of three Qubits realizes an AND gate by choosing the corresponding combinations. Again the output shows a superposition of all possible results of this logical interconnection. The gate has computed the output signal for all possible inputs at once, which is a characteristic for the quantum parallelism. This concept is the basic principle of quantum computers.

The quantum computer uses a fundamentally new principle, which is different from the principle of the Turing machine, the basis of present computers [27]. The inherent parallel processing of quantum computing enables the computation of very complex problems, where classical computers will probably fail completely. As an example the factorizing of large numbers into prime numbers is essential for new concepts in cryptography. Therefore the field of cryptography will become one of the first applications for quantum computing. Once quantum computing is well established many other important applications may come into existence. One of those applications could be an association memory that is a key element for sophisticated speech-recognition systems.

Today the realization of a quantum computer is quite far away although physicists have realized systems with a few Qubits. The difficulty arises from the fact that quantum effects, which usually occur on the atomic level have to be controlled on the macroscopic level.

The realization of Qubits is difficult since they have to form a closed nanosystem without interference effects with the surroundings. In such a system an electron can occupy two states and can interact with a photon so the Qubit can be set. Nowadays ion traps can set the vector of the Qubit with the frequency of the light. An appropriate solution for computers are quantum dots in solid states. However, this concept needs high-purity and perfect materials. Figure 10.22 shows schematically such a quantum-dot field, samples of it have been made and tested in research laboratories.

A very promising concept comes from Mooij from the university of Delft: The Qubit consists of a superconducting ring with three Josephson elements. The magnetic flux quantum with its two states acts as a Qubit, as shown in Fig. 14.13. Advantages are the reasonable coupling of this element to the outside world and the fact that this effect occurs at the macroscopic level.

In principle the quantum computer offers fundamentally new features for data processing on the basis of quantum mechanics without introducing new technologies except those developed for nanoelectronics.

## 5.3 Summary

The examples of the DNA computer and the quantum computer show that different approaches for parallel computing on the nanometer level exist and that they are not restricted to electronics. The difficult problem of efficient parallel processing demands extraordinary concepts, so new innovative concepts will possibly presented in future.

Quantum computing comes up with a fundamentally new concept that goes far beyond the Turing machine. This concept is a good candidate for future applications, such as associative systems or cryptography.

# 6

# Parallel Architectures for Nanosystems

For many years, research has focused on parallel data processing microsystems (Fig. 3.12) in order to enhance the computational power. Therefore, the following discourse discusses to what extent the future nanoelectronic devices are suitable for highly parallel systems. In this context, known architectural concepts have to be checked or modified. This check is necessary, since the new nanoelectronic devices show qualitatively new properties in terms of their physical dimensions, power dissipation, gain, number of stable states, clock frequency, and above all integration level. The integration level is expected to increase from hundreds of million to hundreds of billion devices.

From this point of view, a closer look will be taken at some concepts that will come into play before the DNA computer and quantum computer do. However, the stress lies on the importance of technology for the specific architectural system. This chapter does not claim to examine the fields of parallel data processing in general.

## 6.1 Architectural Principles

### 6.1.1 Mono- and Multiprocessor Systems

Figure 6.1 depicts an overview of mono- and multiprocessor systems. Presently, the most common concept is based on the von Neumann principle and separates the processing unit from the memory (Fig. 6.1a). Usually, the memory is split into a program memory (PM) and a data memory (DM). According to the instructions of the program, the processing unit processes the data. The faster the instructions are executed, the higher will be the performance of the system. When the performance of a single processor is not sufficient, additional processors can be attached to the monoprocessor that, for instance, it evaluates complex algorithms in parallel (Fig. 6.1b).

Today's microprocessors have been successfully designed on this basis. They are advantageous due to their simple digital logic gates and their plain

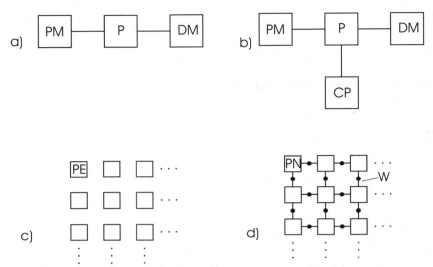

**Fig. 6.1.** Processor architectures (a) processor according to the von Neumann principle, (b) enhanced performance via cooperate processors, (c) processor array, e.g. systolic array, (d) processor nodes PN connected via weights W between the processor nodes

architecture. Their concepts are clear, but anyhow from the point of view of microelectronics today's microprocessors are quite ineffective, since only a small portion of the whole circuit is active simultaneously. The rest of the circuit is in an inactive state and contributes only to the power dissipation. For this reason, future concepts have to shut down these inactive circuit parts.

Another approach tries to process data in a parallel manner with nanosystems. Obviously, the computational power of a system can be extended if many processors work on the same task. In this context, the harmonization of large processor areas turns out to be problematic. Up to now, only a few concepts attain an efficient load balancing. In the long term, a possible solution might exist in a form of relatively loose coupled straightforward devices. In the following, some examples of such concepts will be given. Systolic arrays like in Fig. 6.1c are typical prototypes of today's processor arrays. Connective systems (Fig. 6.1d) like artifical neuronal networks form another processor array type. In this context, the next chapter deals with the relevant concepts.

In the 1980s Seitz illustrated the different concepts in a single diagram for the first time (Fig. 6.2). The diagram shows the number of processor units as a function of the integration level.

Figure 6.2 comprises two extremes: On the left-hand side, memory applications depend only on a few devices per logic function, whereas on the right-hand side, single processor units require all devices of a single chip. In-between, all sorts of combinations are feasible, for instance the associative ma-

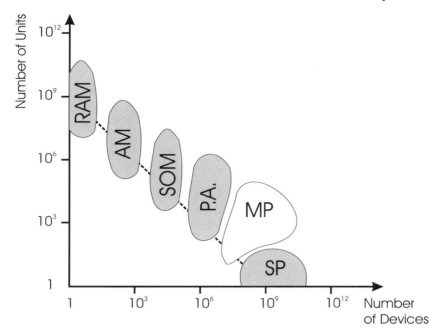

**Fig. 6.2.** Architectures of multiprocessor systems, originally from S. Seitz: programmable logic array (PA), multiprocessor system (MP), single processor system (SP) for a technology level of 1 Gbit memories

trix AM, the selforganizing map SOM, the cellular neuronal network CNN, the systolic array on processor and gate level, as well as dynamic biology-inspired networks. In the following, all these concepts will be discussed from the viewpoint of nanoelectronics.

### 6.1.2 Some Considerations to Parallel Data Processing

Parallel data processing is the most important issue in terms of performance improvements. Therefore, the efficiency of different parallel and timesharing concepts will be analyzed with respect to a single functional block F.

The efficiency $\eta$ is inversely proportional to both the active chip area and the clock frequency. Parallel or timeshared functional blocks do not improve the efficiency, even if we assume that (de-)multiplexing can be neglected (Fig. 6.3). The ratio of throughput and chip area remains constant for both cases.

A single operation $F$ has to be split into suboperations $F_i$, which can be evaluated by parallel subfunctional blocks. At best the efficiency increases according to the number of parallel subfunctional blocks. However, it is unlikely

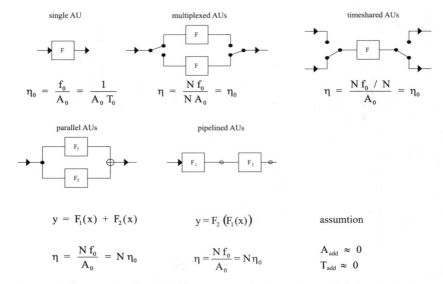

**Fig. 6.3.** Efficiency of different architectures in terms of their parallelism: The upper examples do not offer an enhanced efficiency, whereas the lower concepts improve the efficiency by a factor of N (degree of parallelism) if they comply with the assumption that the additional chip area and switching time can be neglected. According to A. Wieder, F represents the functionality, f stands for the frequency, and A is the chip area

to sum up the subresults without any additional costs. In this context artifical neuronal networks offer a promising concept. The enhanced efficiency of pipeline architectures is based on the space and time parallelism structure. This concept assumes that the operations can be serialized into appropriate suboperations. Such structures can be found for instance, in systolic arrays and fuzzy systems.

The simple configurations in Fig. 6.3 do not reveal the wealth of all possible parallel architectures, but offer a first glance at the additional costs of parallel data processing. Among other structures, typical processor array configurations are, for example: Ring, cylinder, sphere, and hypercube. In general, the decision depends on the software that has to be mapped on the architecture. Additionally, parallel data processing offers the opportunity of a partial fault tolerance on the hardware level.

### 6.1.3 Influence of Delay Time

Nanoelectronics follows the trend to high clock frequencies, since delay and switch times of the tiny devices are short. Clock rates of hundreds of GHz are already considered to be realistic. However, relevant factors tend to be missed

6.1 Architectural Principles 93

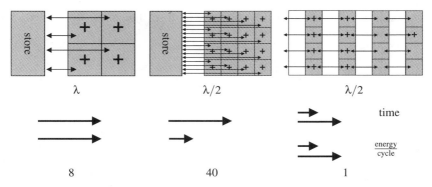

**Fig. 6.4.** Novel computer architectures according to Tiwari [28]: The advantages of scaled parallel systems in terms of power consumption and cycle time. Only distributed computing yields a higher efficiency

at first glance. For instance, the power supply of the clock can consume up to 50% of the whole power consumption. Parasitic wiring capacitances have to be minimized in order to maintain the clock power within certain boundaries. In this connection, a further problem arises: Distributed parasitic capacitances limit the bandwidth of a wire and result in distorted signals.

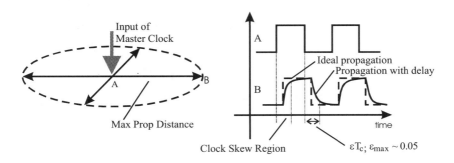

**Fig. 6.5.** The range of the clock signal is limited due to signal delays and distortions

Figure 6.5 illustrates the maximum wiring length. Signal delays and distortions can be tolerated within this limit. In this context, the estimation of Fig. 6.6 takes into account the wiring dimensions and the clock frequency. It turns out that only small areas can be reached from a single point. However, the packing density of nanodevices is very high so that many devices can still be accessed.

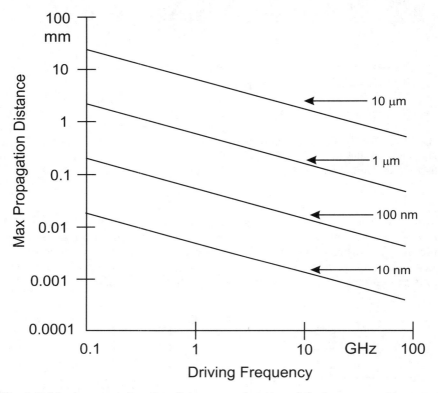

**Fig. 6.6.** Maximum propagation distance as a function of the frequency with respect to the wire dimensions. M. Forshaw from UCL

As is known, networks on the circuit level are sensitive to timing delays, but, nevertheless, the same is true for networks on the processor level. For instance, Fig. 6.7 illustrates a systolic processor array of 3 × 3 processors. Typically, the results do not appear at the same time, but emerge with a delay of two clock cycles. Thus, new operations can only be initiated at each second clock cycle and the processing speed gets limited by the time delay.

Appropriate delay elements in the data paths omit this drawback. By this means, all results leave the processor array simultaneously and can be processed in a next step without any further delay. Therefore, delay times of the wiring are not an issue within such configurations. However, the result gets delayed.

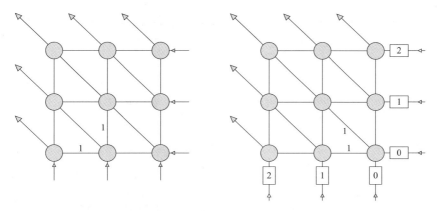

**Fig. 6.7.** The impact of time delays can be cut down to the processing speed via delay elements

### 6.1.4 Power Dissipation and Parallelism

The extremes of parallel data processing will be discussed on the base of the SIMD architecture (Sect. 6.2.3). In this context, certain considerations of the MOS technology can also be applied to nanoelectronics. The degree of parallelism $P_P$ can be expressed as:

$$P_P = \frac{N_{PE}}{N_{OP}}. \tag{6.1}$$

$N_{OP}$ represents the number of operations of the single processor, whereas $N_{PE}$ stands for the number of parallel processors. If one processor takes the time $t_0$ for the evaluation of one operation, the overall system cuts this time down to:

$$t_{ds} = \frac{t_0}{P_P}. \tag{6.2}$$

Thus, the higher the degree of parallelism, the faster the evaluation of the task. This is of fundamental importance for all time-critical tasks.

In terms of power consumption other considerations have to be taken into account. The time $t_{ds}$ has not to be reduced, and the time $t_0$ can be enlarged. The total power consumption comprises a dynamic and a static part:

$$P_{tot} = P_{dyn} + P_{stat}. \tag{6.3}$$

The dynamic part results mainly from the load capacitance $C_L$:

$$P_{dyn} = C_L V_{DD}^2 f_c. \tag{6.4}$$

In this context, $V_{DD}$ refers to the power-supply voltage and $f_c$ stands for the clock frequency that is inversely proportional to $t_0$:

$$f_c = \frac{C_C}{t_0}. \tag{6.5}$$

$C_C$ is the accompanying proportionality factor. It follows from the above equations for the dynamic power consumption:

$$P_{dyn} = C_{LPE} V_{DD}^2 P_P N_{OP} \frac{C_C}{t_{DS} P_P}. \tag{6.6}$$

The dynamic power consumption does not depend on the degree of parallelism if all parameters remain constant. However, it can be significantly reduced if the supply voltage is diminished. Though, this results in a slower gate switching which is no issue as long as the time $t_{DS}$ does not increase.

The switching time of a CMOS gate can be approximated by:

$$t_{dI} = \frac{C_{LI} V_{DD}}{\beta (V_{DD} - V_T)^\alpha}. \tag{6.7}$$

$C_{LI}$ represents the load capacitance, $\beta$ stands for the technology constant and $\alpha$ (usually in the range of 1 to 2) accounts for the saturation effect of the carrier mobility. Assuming that a single operation has to pass through $n_{PE}$ gates within one processor element, one can state with respect to (6.2):

$$n_{PE} \frac{C_{LI} V_{DD}}{\beta (V_{DD} - V_T)^\alpha} \leq t_{ds} P_P. \tag{6.8}$$

If the application demands a specific processing time $t_{ds}$, $V_{DD}$ can be expressed as a function of $P_P$ ($\alpha = 2$):

$$V_{DD} = V_T + \frac{b}{2P_P} + \sqrt{\frac{bV_T}{P_P} + \frac{b^2}{4P_P^2}}. \tag{6.9}$$

In (6.9) b is equivalent to $n_{PE} C_{LI}/t_{ds} \beta$. Thus, a constant $t_{ds}$ and a higher degree of parallelism results in a reduced $V_{DD}$. However, $V_{DD}$ can only be reduced as far as the transistors enter the subthreshold domain (Chap. 3), since the circuitry gets too slow in this domain.

The static power consumption increases for a constant supply voltage with the degree of parallelism, since a larger number of devices cause more leakage current.

These reflections prove the existence of an optimum degree of parallelism in terms of the overall power consumption $P_{tot}$. The overall power consumption is dominated by its dynamic portion if the degree of parallelism is low because of the high supply voltage, whereas it is ruled by its static part if the degree of parallelism is high. The same is true for nanoelectronics, except that the static part might be neglected.

## 6.2 Architectures for Parallel Processing in Nanosystems

The following integratable concepts reveal the feasibility and problems of nanosystems.

### 6.2.1 Classic Systolic Arrays

A known example is the systolic array. It is base on a regular structure (Fig. 6.8) and data gets pumped through this structure in a way comparable to blood circulation. Kung and Leiseron suggested systolic array algorithms that are based on modular design principles and exploit the spatial and temporal parallelism of data. The aim is to solve a given problem with p parallel processors, which should result in a speed up of p in comparison to a single processor.

The literature already comprises many areas of systolic-array applications. The most important ones are: Linear programming (vector and matrix operations), convolution systems, solving of partial differential equation, numeric (solving nonlinear differential equation, linear and nonlinear optimization, etc.) and graph algorithms. The name 'systolic array' accounts for their rhythmic interaction between processor action and data transport. The area/time product serves as a benchmark for the modular systolic structures. The following example reveals the timing and communication challenges of a vector/matrix multiplication array.

The matrix $\underline{W}$ with its components $w_{11} \cdots w_{33}$ has to be multiplied by the input vector $X$ (Fig. 6.8a). The output vector $Y$ results from $y_1 = w_{11} \cdot x_1 + w_{12} \cdot x_2 + w_{13} \cdot x_3$, $y_2 = w_{21} \cdot x_1 + w_{22} \cdot x_2 + w_{23} \cdot x_3$ and $y_3 = w_{31} \cdot x_1 + w_{32} \cdot x_2 + w_{w3} \cdot x_3$. First, each component $w_{ij}$ of the matrix $\underline{W}$ gets stored in a single processor element. Then the input vector is pumped through the processor array. During the first clock cycle, the input element $x_1$ is multiplied by the element $w_{11}$ (Fig. 6.8b), summed up with the value on its right side and finally shifted to the left. In order to use the same design for all processor elements, zeros are inserted from the left side. During the second clock cycle $w_{12}$ is multiplied by $x_2$, gets added to $w_{11} \cdot x_1$, which comes from the left side and conclusively is shifted to the right side. At the same time, in the second row $w_{21}$ is multiplied by $x_1$ and so on (Fig. 6.8c - f). After the third clock cycle the first component $y_1$ is available and after the fifth clock cycle the complete output vector $Y$ is available. The next input vector $Z$ has to follow directly after $X$ so that each clock cycle delivers a new output vector. This means, for instance, that $z_1$ and $x_2$ have to be inserted simultaneously.

This example evaluates 27 multiplications of a vector - matrix multiplication. A single processor can only evaluate these multiplications in a serial fashion, whereas 27 processors would be needed for a parallel solution. The advantage of the systolic array is based on the fact that only three processors can process the weights $W$ and the input value $X$ in a very resourceful way.

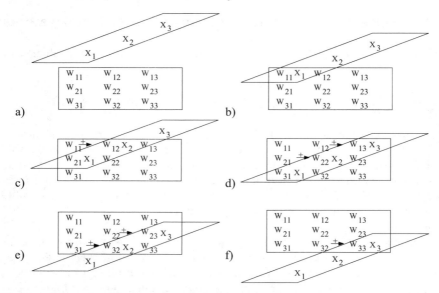

**Fig. 6.8.** Example of the organization and function of a systolic array: matrix/vector multiplication as it is applied, for instance, in neuronal networks

Taking a) into account, only six individual steps are needed, which leads to an area of efficient and fast implementation.

### 6.2.2 Processors with Large Memories

In the past, new processor and memory concepts led to higher performance: Within the same technology and at the same clock frequency, today's RISCs (reduced instruction set computers) typically show a higher performance than today's CISCs (complex instruction set computers). Typically, the performance is about a factor of 2 to 4 higher. The circuitry of a RISC is so simple that it can be integrated on a fraction of the chip area of a CISC. Thus, the production as well as the design of a RISC is less expensive and time consuming. This is why most of today's processors are based on a RISC concept.

The steady integration improvements have increased the ratio of the microprocessors and the memories speed. The drawbacks of the bottleneck between the processor and memory can be bypassed with a very large on-chip memory. Figure 6.9 shows the face plan of the IRAM (intelligent RAM) as an example of such a structure. The memory consumes roughly 80% of the total chip area whereas the remaining part is reserved for the RISC processor with an additional vector processor that serves for multimedia purposes.

The maximum clock frequency can be roughly estimated as a function of the address space and the memory size. The bigger the address space the more time consuming the operation, e.g. an addition in the processor. The bigger

## 6.2 Architectures for Parallel Processing in Nanosystems

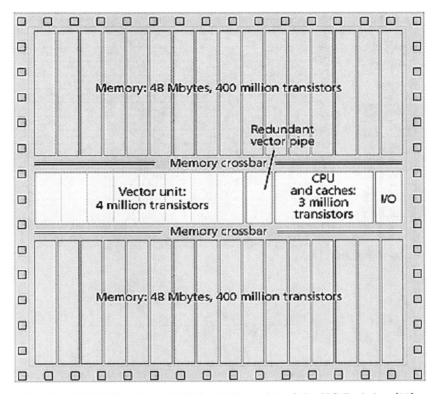

**Fig. 6.9.** IRAM (inteligent RAM), E. Kozyrakis of the UC Berkeley, [29]

the memory capacitance the more extensive the memory itself and the greater the delay time.

A further performance increase is attainable via multiport memories [30]. A single-port memory system consumes three clock cycles for one addition: The operand is latched in a register, then the second operand is read and the addition takes place and finally the result is stored. In terms of a three-port memory all these memory operations can take place simultaneously so that the addition operation only consumes one clock cycle. Thus, the system speeds up by a factor of three. A six-port memory in connection with an additional processor results in a further speed up of about a factor of two. Obviously, this method enhances the computation performance, e.g. a thirty-port memory causes a speed up of about one order of magnitude. This concept of multiport memories might also be interesting for nanoelectronics.

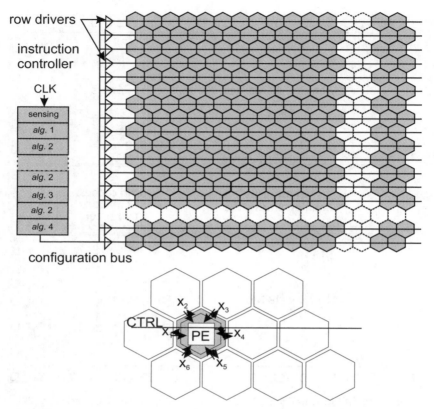

**Fig. 6.10.** SIMD architecture of a fingerprint recognition system, S. Jung [31]

### 6.2.3 Processor Array with SIMD and PIP Architecture

Multimedia applications are the objective of the IRAM architecture, the same is true for SIMD (single instruction stream, multiple data stream computer) processor arrays [32]. For instance, the fingerprint recognition system in Fig. 6.10 is based on a SIMD architecture. The same global instructions are fed into all processor elements that handle, for example, image-processing data. The global supply of all processor elements with the same instruction is a challenging issue for nanoelectronics, because of the extensive wiring. Apart from the global wiring and due to the regular structures and interconnects, the SIMD structure offers an ideal architecture for nanoelectronic-based system integration.

The problematic wiring loses its tension for the PIP (propagated instruction processor) (Fig. 6.11). The PIP evolves from the SIMD concept. Within the PIP the operations are pumped through the network, while the PIP elements evaluate their operations. According to this idea, the network can evaluate different algorithms simultaneously.

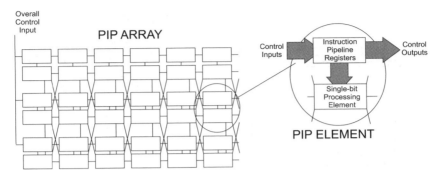

**Fig. 6.11.** PIP (propagated instruction processor) architecture

### 6.2.4 Reconfigurable Computer

The RAW (reconfigurable architecture workstation) machine in Fig. 6.12 consists of a repetitive architecture. Each processing element comprises a local instruction unit and data memory. The complexity of a single processor element is equivalent to a medium-range microprocessor and roughly comprises two million transistors. The logic function of each processing element can be programmed via reconfigurable logic blocks, which is very important for nanoelectronics. Furthermore, programmable matrix switches permit point-to-point connections. The programmable connections of this concrete example consume about 30% of the whole chip area.

Besides the valuable regular structure, the RAW machine also offers the possibility to map distinct hardware tasks into software, which generally is advantageous for implementation purposes. This concept is also capable of error-tolerant implementations, since reconfiguration and rewiring information can be stored in a program that might be based on benchmark data. A further substantial increase of the computational performance results from the localized reconfigurability.

Figure 6.13 reveals the different levels of reconfigurable circuitries: Logic level, data-path level, arithmetic level, and control level.

### 6.2.5 The Teramac Concept as a Prototype

Probably, error-tolerant architectures have to be assumed in order to produce forthcoming systems comprising billions of devices in an economical fashion. Every computer that is based on nanoelectronics will include a considerable

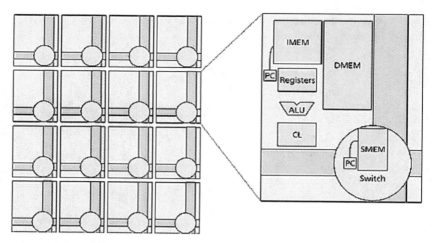

**Fig. 6.12.** RAW (reconfigurable architecture workstation) machine based on a regular and flexible structure

number of defective devices, wires, and switches. Such a concept differs considerably from the ordinary complex computer concept. In this context, the Teramac concept should be mentioned as a prototype concept that has been built for research purposes. It is a user-configurable parallel computer machine that comprises many defects. However, it not only works, but also has a very high computational power. The concept is an inexpensive approach that is not based on an error-free hardware platform. First, the hardware localizes all of its defects and configures the parts that are in working order. Then, the program has to be compiled before it can be executed. Such concepts are based on an excessive wiring structure, which can be established by fat tree-structured networks (Fig. 6.14). On this basis, significant knowledge could already be gathered with regard to nanotechnology [33].

The Teramac comprises $10^6$ logic units that are clocked at a relatively low frequency of $1\,MHz$. This is equivalent to $10^{12}$ operations per second, i.e. tera operations per second. This feature is responsible for the Teramac's name. The Teramac system has been realized in today's microelectronic devices and comprises six levels. The lowest level performs the data processing by means of look-up tables (LUT). The upper levels connect together the LUTs and form more complex functional blocks. The wiring structures are designed in a highly redundant fashion. Programmable switches attain an enhanced flexibility.

A first conclusion from the Teramac experiment claims to build an efficient computer as long as sufficient bandwidth is available to find functional components. A second conclusion states that the functional blocks of a computer need not be arranged in a regular structure, but have to have a complex wiring scheme with a kind of technical intelligence. Only by this means can defective

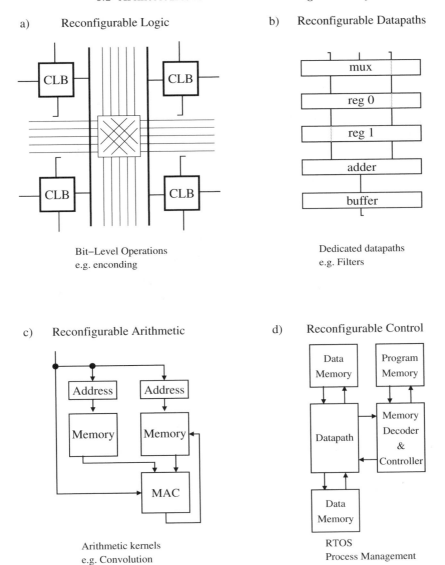

**Fig. 6.13.** Different methods for reconfigurable circuitries, J. Rabaey, UC Berkeley

104    6 Parallel Architectures for Nanosystems

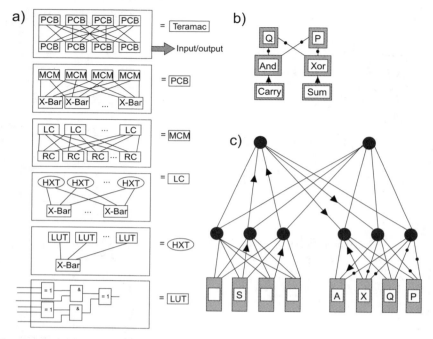

**Fig. 6.14.** Architecture of the Teramac: It is based on a hierarchical structure with a high degree of redundancy (a). The right side (b and c) reveals the implementation principle of logic functions

elements be eliminated in an effective way. The third conclusion concerns the switching devices: The most important components of future nanoelectronic circuitries will be the address bus and the matrix switches, since they rule the hardware configuration (Fig. 6.15). The new paradigm claims that difficult hardware tasks should be scheduled as software tasks. This is equivalent to the development of the microprocessor, namely to put complex application programs into software.

The conclusions of the Teramac experiment offer complete new concepts for nanoelectronic research: It is a top-down concept and not a technology driven bottom-up concept. Key components are wiring, switches, and memories as they guarantee efficient nanoelectronic structures.

## 6.3 Summary

The steady improvements of microelectronics result in a continuous increase of computational power in terms of the classic processors. On the basis of already gathered experiences one can claim that the simple von Neumann processor

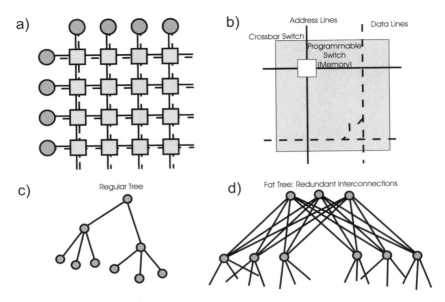

**Fig. 6.15.** Configuration of the interconnection points (a and b) and the idea of fat tree-structured networks (c and d)

structure will retain its importance. New system architectures for parallel data processing are under intensive investigations. This is where microelectronics comes in, because only by means of an appropriate hardware implementation can the advantages of, e.g. multiprocessor structures and extensive wiring structures, be exploited. The advantages of the hardware will increase further when it comes to nanoelectronics.

# 7
# Softcomputing and Nanoelectronics

Artificial neural networks (ANN), fuzzy systems FS), and evolutionary algorithms (EA) or genetic algorithms (GA) provide interesting approaches for nanoelectronic architectures. These three concepts are summarized by the term computational intelligence (CI), or softcomputing. They will extend the capabilities of conventional computer systems, because softcomputing facilitates a better data acquisition from our everyday environment [34]. CI methodologies will enable nanoelectronic systems with new and very special characteristics [35].

By applying the concept of fuzzy logic to an information-processing system the fuzzy controller can be realized. This controller is based on linguistic rules, so that human experts can insert knowledge into the system. The behavior of

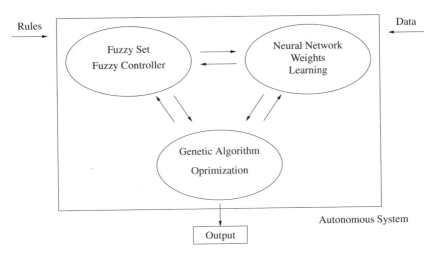

**Fig. 7.1.** Schematic diagram of an autonomous system based on computational intelligence methods

108     7 Softcomputing and Nanoelectronics

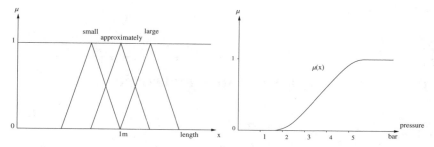

**Fig. 7.2.** Examples of membership functions

neural networks is adapted by input data supplied by the environment. These networks learn by adjusting the strength of connections (weights) between the neurons. Although the entire system turns out to be very complex, it can be optimized by applying genetic algorithms. By using these three methods, a completely autonomous system can be realized (Fig. 7.1). Autonomy is an important feature for nanoelectronic components, because programming these systems from the outside will be a serious problem [36].

## 7.1 Methods of Softcomputing

### 7.1.1 Fuzzy Systems

Fuzzy systems offer two important advantages for very complex nanoelectronic systems: First, the structuring and programming of these systems can be carried out by using linguistic data. Secondly, the processing of fuzzy values is very robust to variations of absolute quantities, which is an important feature for nanoelectronic switching elements. The potentialities of the second advantage are barely investigated as yet.

The concept of fuzzy or unsharp values is very familiar to nanoelectronic considerations. Heisenberg's uncertainty principle and the probabilistic meaning of Schrödinger's equation are fundamental facts. By using these equations we can calculate the probability of finding a particle in a specific spatial region, as described in Chap. 3.

But the fuzzy theories meaning of unsharpness is different from the consideration stated above. In this case we have to distinguish between randomly distributed and linguistic unsharpness. Random unsharpness is modeled by the probability theory. Linguistic unsharpness describes the uncertainty of a textual characteristic, for example fast, big, small, etc. The probability theory requires that all possible events of an experiment are known and that the sum of independent probabilities equals one. This does not apply to linguistic unsharpness, which is an important difference.

Today fuzzy logic is used as a common expression for the theory of fuzzy sets and the fuzzy logic itself. The theory of fuzzy sets can be regarded as an

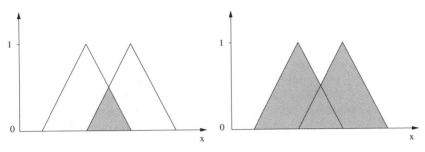

**Fig. 7.3.** Fuzzy operations: Minimum and maximum operation

extension of the classical set theory, which only classifies elements in matching and nonmatching types. This concept of matching is extended for the theory of fuzzy sets. The factor $\mu$ indicates the degree to which an element matches a particular set. Thereby 0 and 1 correspond to the classical case of matching and nonmatching. For the matching degrees between 0 and 1 the element only belongs partially to a specific set. Figure 7.2 presents fuzzy sets as mathematical descriptions of linguistic relativization of quantities, for example a "high" pressure or a length of "about 1 m", "a little shorter than 1 m", or "a little longer than 1 m". The depicted triangular functions are arbitrarily chosen. Normal distributions are alternatively used to define fuzzy sets, because they can be completely described by two parameters. These membership functions contain the expert knowledge of a fuzzy system. They must be varied in such a way that the given task is solved as well as possible.

The set concept extension to fuzzy sets has the advantage of coming closer to the human understanding than the classical set theory. It enables us to model complex and almost incomprehensible systems. Corresponding to the examples in Fig. 7.2 we can define membership functions for any object or state that can be described or characterized by a human verbalism. This process of assigning a membership function to an object or state is called fuzzyfication.

Comparable to conventional digital logic we also have to define a set of basic operations for fuzzy sets. First, these operations are intersection (minimum) and conjunction (maximum), which are illustrated in Fig. 7.3.

Computers are not only suitable as data-processing machines, but also as knowledge-based systems. They can process data as well as knowledge in the form of rules. To make this practicable we have to implement an intensive networked set of switching elements, which results in high demands on the system architecture. One realization of such a system is the fuzzy controller, which is briefly explained in the following:

By introducing the fuzzy technique we can feed linguistic data into a computer. For this purpose, the fuzzy controller was developed. Today this controller is mainly employed in control engineering, which led to considerable improvements in process-control applications. What kind of significant effects does this concept have on nanoelectronics? The field of application is mainly given by autonomous systems. Because the number of connecting terminals for large-scale systems is always restricted, we have to apply concepts that can structure and program themselves.

At first we will show how rules that occur in knowledge-based systems can be explicitly represented by fuzzy relations. A fundamental characteristic of fuzzy logic is the linguistic, often unsharp presentation of quantities. This unsharpness should not be interpreted as inaccuracy, because fuzzy logic is absolutely capable of producing exact results. Fuzzy logic makes high demands on microelectronics, but also presents promising opportunities.

For complex tasks with many input variables several rules are needed, whereby a premise may consist of many partial premises. For this purpose Mamdani developed the following evaluation strategy which is illustrated in Fig. 7.4:

- If a premise consists of several partial premises, the truth value is equivalent to the lowest truth value of the partial premises.
- The conclusion is limited to the rules truth value.
- If several rules produce a truth value $> 0$, then the rule conclusions will be superposed.

For the control of real processes it is often necessary to convert the fuzzy set into a discrete correcting variable, which is called defuzzyfication. Therefore several methods exist, which were mainly proposed intuitively. The most common technique is the center-of-gravity method, which is also illustrated in Fig. 7.4.

The rules shown in Fig. 7.4 apply to the simple example of charging a NiCd battery. A typical effect of this application is the aging of the battery cell, which can not be described mathematically but can be covered with linguistic rules of fuzzy logic. The input variables, which correspond to measured values, are presented as so-called singletons. The appropriate output value is calculated from the intersection sets by a minimum operation, which is limited by the truth value of the respective rule. The results of the individual rules are combined by a maximum operation. By using the center-of-gravity method, a discrete value is derived from the resulting fuzzy set. In this context the fuzzy controller is not described in too much detail. We will only treat the fundamental aspects that are relevant to microelectronic applications.

In this simple case the output signal's dependence on the two input variables can be illustrated by the three-dimensional plot in Fig. 7.5. The charging plot produced by the fuzzy controller is single-valued and absolutely not unsharp. The fuzzy controller is a nonlinear analog operational unit, which can be compared to an approximation system. It produces a control signal to near-optimally charge a battery by applying the linguistic set of rules. If a complete mathematical model of the battery existed, the exact charging control characteristic could be calculated. But although this is not the case, the problem can still be solved approximately by using a fuzzy controller with specific linguistic rules.

## 7.1 Methods of Softcomputing

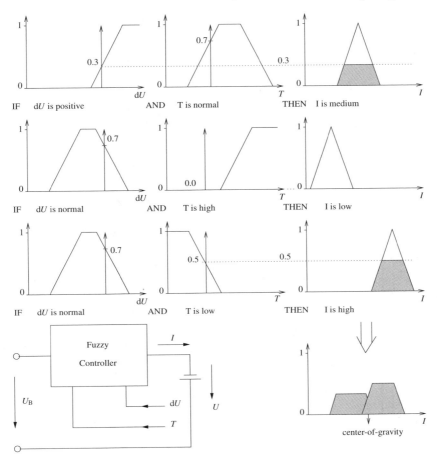

**Fig. 7.4.** Illustration of fuzzy-system rules for the example of charging an NiCd battery

According to microelectronics two aspects are very important: First, the control characteristic has to be described by a small set of rules determined by a simulator. It can be shown in practice that often less than five rules are sufficient. Moreover, the type of membership function should be determined as simply as possible. Both steps are suitable to reduce the circuit implementation effort drastically. Secondly, only the principal characteristic of the charge control is needed. That is why relatively inaccurate fuzzy-gate functions, which can be realized with less effort, are sufficient to build a controller. This strategy becomes especially important, if we want to develop more complex systems, for example, a complete array of image processing fuzzy elements built from sub-$\mu m$ transistors.

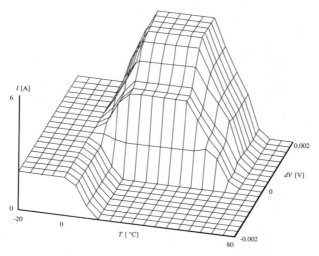

**Fig. 7.5.** The charge current $I$ according to the temperature $T$, and the voltage change $dV$. The nonlinear, analog output function can be constructed by the given fuzzy rules

A fundamental characteristic of fuzzy rules is a probably accurate result, although the computation is based on uncertain data. This feature can be of great significance for nanoelectronic applications. Because of the very fine structured devices and because only a few particles are involved in data processing, significant parameter variations may occur. These variations also lead to fluctuations in output signals. Therefore techniques have to be developed that take these circumstances into account.

### 7.1.2 Evolutionary Algorithms

The transfer of evolution processes to technical applications is an interesting task in information theory. Comparable to artificial neural networks and fuzzy systems, genetic algorithms are inspired by biological concepts. In analogy to an organism adapting to a changing environment by evolution, a technical system could adjust itself to changing boundary conditions by artificial evolution (Fig. 7.7). For this purpose the system is represented by a data set that is exposed to mutation and crossover operations. The resulting data set is maintained if it is improved compared to the original set, otherwise it will be ignored (Fig. 7.6).

This decision demands a target function: The survivability of an organism is the main goal of biological processes. However, the priority task of technical systems is the optimization of special characteristics, for example, their power or hardware effort. The evolutionary methodology is only useful for very complex systems that cannot be mathematically modeled, which definitely holds true for nanoelectronic systems.

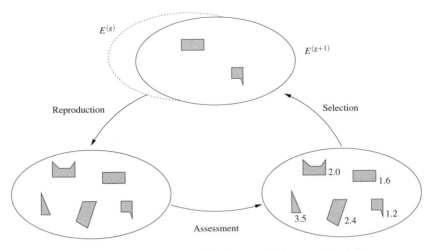

**Fig. 7.6.** The basic principle of an evolutionary algorithm

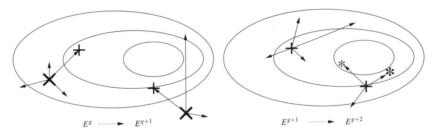

**Fig. 7.7.** The selection mechanism of an evolutionary algorithm. The picture illustrates two iterations for the minimization of a two-dimensional potential function

The method of genetic processes in applications did not arise until fast computer systems were available, which enabled the computation of artificial evolution. A major goal of nanoelectronics will be the implementation of evolutionary systems in nanoelectronics itself. Therefore evolutionary algorithms have to be parallelized and efficiently implemented in electronics. Thereby the problem of wiring will play an important role, because the data sets described above have to be read, altered, and transmitted by the system.

### 7.1.3 Connectionistic Systems

Connectionistic systems represent a very promising approach to processor arrays, which could solve some common problems of parallel processing. Neighboring processors of such arrays are directly connected. The connection strengths (weights) determine the behavior of the net. The weights depicted in Fig 7.8 represent the program stored inside the net, comparable to a neuron illustrated in Fig. 7.9. On the one hand, the connections to the weights have to

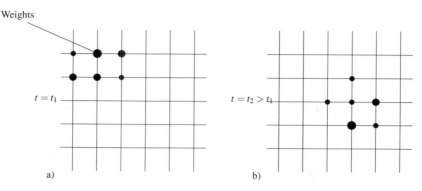

**Fig. 7.8.** A net of parallel processors, where the weights determine the connections between the processing elements. Two time-varying states are depicted

be drawn, on the other hand the weights have to be loaded. The development and adaption of weights in biological systems is based on heredity (transmission) and learning. For technical systems a similar approach is applied: First, we try to calculate appropriate weights by simulation. Then, the computed values may be implemented. The finetuning of the weights is carried out subsequently by learning, which can be a time-consuming process. From the microelectronic point of view such arrays exhibit interesting features, namely association, self-organization, and fault tolerance.

A special case of connectionistic systems are neural networks, which are derived from the biological nets of nerves, where all neurons may be connected amongst each other. In the following we will present four examples of neural networks. For each of them an implementation idea for VLSI systems exists. From a VLSI point of view they have the advantage of modularity and fault tolerance.

Artificial neural networks are an important challenge for nanoelectronic implementations. First, large networks with many parallel processing neurons are needed for relevant applications. Secondly, neural networks have characteristics that are also important to nanoelectronics and VLSI CMOS circuits, for example fault tolerance and the ability of adaption. The basic principle of a simple neural network is illustrated in Fig. 7.9, which shows the schematic structure of a single neuron. The neuron weights that are determined by the modified Hebb rule (described in Sect. 4.1.1) are stored in the depicted crossing points. The products of weights and input signals are added and transmitted over an additional signal line and the resulting sum is compared to the threshold value $T_h$. If the signal reaches the threshold value the output is set to one.

Figure 7.10 illustrates a two-input neuron with its characteristic input - output surface. The defined weights divide this space into two sections. If we extend the examined network by a second level of neurons four different sections will arise in the output space. This network configuration can solve the

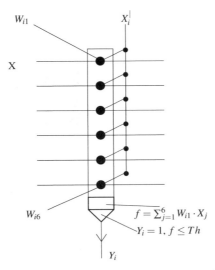

**Fig. 7.9.** An example of the structure of an artificial neuron

XOR problem, which was considered to be impossible by using only neurons until the 1960s. If we further extend the network by a third processing level, every possible segmentation can be constructed and the network can be used as a universal mapping system. The main implementation problem of these networks is the task of learning. As long as no solution to this problem was known, neural networks were considered to be useless. Below we will present some learning methods that became established.

The artificial neuron presented in Fig. 7.9 and 7.10 is very similar to a threshold gate. The weights determining the thresholds that are adapted by learning are the main difference between these two approaches. Concerning threshold gates, these weights have to be predetermined in an optimal form.

A special case of neural networks are cellular networks, which are characterized by the strictly local type of neuron connections, i.e. only direct neighbors are linked. Cellular networks try to globally solve a given task only by local information processing. The local connections may consist of simple weights or so-called fuzzy links.

### 7.1.4 Computational Intelligence Systems

Figure 7.11 shows some applications of autonomous softcomputing systems [37]. The number of synapses, which equals the number of weights, indicates the level of integration. A measure of the networks performance is given by the number of connections updated per second (CUPS). In practice, today's applications are implemented with VLSI devices. For the realization of complex problems, e.g. image processing for autonomous control of vehicles, more sophisticated technologies such as $\mu m$-CMOS are needed.

# 7 Softcomputing and Nanoelectronics

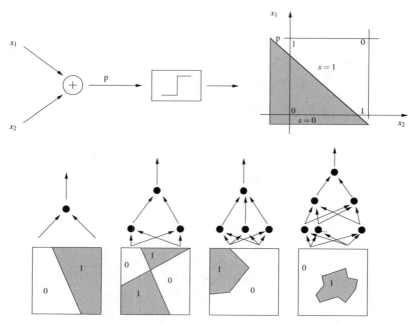

**Fig. 7.10.** The main function of one neuron and of multilayer neural networks

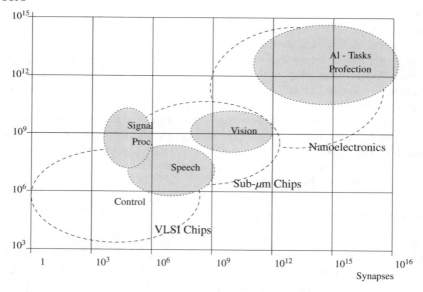

**Fig. 7.11.** Application areas for computational intelligence networks

The hardware requirements for real intelligent and autonomous systems are several orders of magnitude higher. Many fast conventional processors cannot solve these tasks sufficiently, because the processing could not be done in real time and because the volume of such processor arrays is too large. Only networked nanoelectronic systems with a high degree of parallelism offer a promising way to implement such complex applications.

## 7.2 Characteristics of Neural Networks in Nanoelectronics

In order to handle high complexity problems new features have to be implemented in nanoelectronic systems. Such features are local processing, fault tolerance, distributed storage, and self-organization.

### 7.2.1 Local Processing

Cellular neural networks (CNN) are an example of local processing. These simple neural networks were developed by Chua from cellular automata. The processors of such networks are only locally connected, whereby the weights are often considered as binary values (Fig. 7.12). The basic algorithm is principally given by:

$$V(i,j,n+1) = f\left\{ \sum_{k=-1}^{+1} \sum_{l=-1}^{+1} A(k,l)V(i+k,j+l,n) \right. \\ \left. + \sum_{k=-1}^{+1} \sum_{l=-1}^{+1} B(k,l)E(i+k,J+l) + I \right\}. \quad (7.1)$$

$V(i,j,n)$ is the output value of the cell $C(i,j)$ for the time step $n$. $E(i,j)$ equals the input value of the cell $C(i,j)$. $A$ and $B$ are operators, $f$ is a nonlinear function, and $I$ is a bias value. The principal function of a cell and its network integration is shown in Fig. 7.12.

An application is a photodiode-matrix that delivers the input signals of a cellular network. To reduce the noise of a picture that was taken by the matrix, we have to apply the following operators:

$$A = \begin{pmatrix} 0 & 1 & 0 \\ 1 & 2 & 1 \\ 0 & 1 & 0 \end{pmatrix} \quad and \quad B = \begin{pmatrix} 0 & 0 & 0 \\ 0 & 0 & 0 \\ 0 & 0 & 0 \end{pmatrix}. \quad (7.2)$$

The values of a cell itself and the values of the neighboring cells are assessed by the operator $A$. The input signals, for example, the light signals scanned by the photodiodes, are adapted by the operator $B$. The results of this operation are added and increased by the bias $I$. Then this sum is assessed by the nonlinear

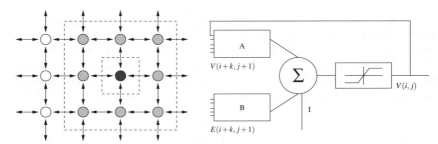

**Fig. 7.12.** The architecture of a cellular network and the basic principle of a single cell

function $f$, which results in the output signal $V$ of this cell. Comparable to this noise-reduction algorithm, other templates are known that are commonly applied to image processing tasks, e.g. edge detection, structural correction, etc.

Because of their structural simplicity and because it is very difficult to completely connect all processors of a nanometer-scale device among each other, cellular networks are a convenient concept for nanoelectronic or molecular-electronic system implementations. On the other hand, the information-processing capabilities of cellular networks are restricted, which means that this concept is limited in its practical importance.

### 7.2.2 Distributed and Fault-Tolerant Storage

An associative matrix is a very simple neural network. It is a memory device that is capable of storing associative data words. When applying the first data word as an input, the best matching data word is read from the memory matrix and transmitted to the output. Associative and distributed storage are interesting concepts from the nanoelectronic viewpoint. Both ideas will be explained by the example of a simple and small associative memory.

Today the data stored inside conventional microelectronic memory circuits can only be recalled by its address or by its content. Based upon artificial neural networks a distributed storage, where data is saved spread over a matrix, can be carried out.

An associative matrix is based on the same principle. The weights that are stored on the matrix points can only attain binary values, i.e. zero or one. In the learning phase we have to apply the associative data pairs to the matrix input and output. The learning algorithm will set the weight of a matrix intersection point to one, if both the input and the output according to this intersection are set to one. The matrix values are subsequently adapted by supervised learning. Figure 7.13 shows for a simple example, how data patterns are saved on the matrix. In order to read data from the matrix an input pattern is at first multiplied by the stored weights. Then the products of

## 7.2 Characteristics of Neural Networks in Nanoelectronics

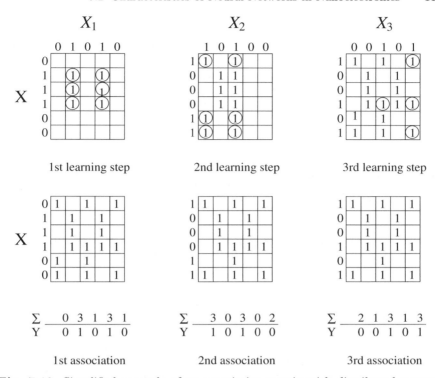

**Fig. 7.13.** Simplified example of an associative matrix with distributed storage. The matrix consists of 5 neurons with 5 inputs each. The weight can only attain the values 0 and 1

each matrix column are added. If the resulting sum is higher than a threshold value, the output is set to one, otherwise to zero. In this way the output data word for a given input pattern can be reliable associated.

This principal functionality can only be maintained if we assume a sparsely occupied memory matrix. If all matrix positions contain a one, the association algorithm would fail and the memory matrix would be useless. In contrast to other common microelectronic memory devices the associative data is stored distributed all over the matrix. Palm [38] derived equations for the calculation of associative memory capacities and showed that such devices may hold 96% of the capacity of classical devices. He also deduced an expression for the unavoidable rate of data errors that may occur for high degrees of matrix occupation. This concept is only relevant for Gbit associative memory matrices.

If some of the matrix memory cells fail, the association process may still work as expected, i.e. the network is fault tolerant to some degree (Fig. 7.14). To achieve a robust fault tolerance the thresholds have to be slightly decreased, which in turn also leads to a decreased noise immunity. This also means that an

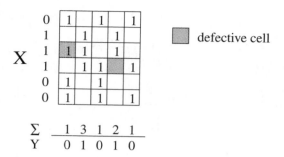

**Fig. 7.14.** Association and hardware errors in the matrix. The shaded cells represent defect cells in comparison to Fig. 7.13. If the threshold is decreased to 2 the results are still correct

upcoming system failure can be detected early (graceful degradation). Fault-tolerance properties will gain further importance in future electronic systems, because the complete functionality of those system components cannot be guaranteed.

### 7.2.3 Self-Organization

An additional characteristic property of neural networks is the ability of self-organization. A neural network can automatically adjust its weights according to the presented input data, so that a suitable mapping to the network is achieved. This property is not only interesting for information-processing purposes. It is also an important aspect for nanoelectronics, where homogeneous arrays of many small elements have to be structured for the desired behavior of the whole system.

In 1978 a model that could explain the self-organization processes in the human brain was searched for, and Kohonen invented the biology-inspired algorithm of self-organizing maps (SOM) [39]. The structure of such a map is given by an array of processing elements (neurons) that are connected among each other. The $n$-dimensional input vectors, which are subsequently applied to the input of this network, are automatically mapped to their specific best-matching location on the two-dimensional map plane. That is how an $n$-dimensional input space is mapped by self-organization on a two-dimensional network plane.

In the self-organizing maps learning phase a restricted set of correlated-data vectors is applied in a statistically distributed sequence to the processing elements of the map. For every given input pattern the map is searched for the so-called best matching unit (BMU). This unit is the processing element, which carries the reference vector that is most similar to the given input vector. After estimating the position of the BMU an adaption takes place, which affects all processing elements in a given range to the best-match position. This range, also called the neighborhood radius, is subsequently decreased

## 7.2 Characteristics of Neural Networks in Nanoelectronics

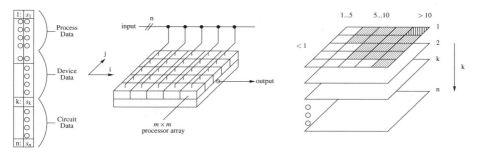

**Fig. 7.15.** Basic principle of a self-organizing map

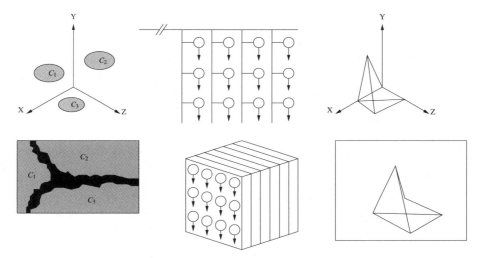

**Fig. 7.16.** Mapping a three-dimensional data space onto a two-dimensional map

during the learning process. That is why the map is roughly structured in the beginning of the learning phase, and fine tuned at the last learning steps. The disadvantage of this very effective parallel algorithm is the very large number of neuron interconnections on the map.

At the end of the learning phase all processing elements have converged to specific values, i.e. adapted reference vectors. The resulting map can be divided into $n$ layers, which represent the components of the input-data vectors. These layers are often presented as a grey-scale image (Fig. 7.15), in order to visualize the mapped distribution of a specific component from the input-data space.

This self-organized network maps $n$-dimensional input patterns onto a two-dimensional plane (Fig. 7.16). The process of self-organization relates similar input-data vectors to network elements that are close to each other. This topology-conserving effect enables us to analyze component correlations by comparing the different map layers. The self-organizing maps are today used for data analysis, classification, and prediction tasks. On a trial base they are being applied for microelectronic applications in process control and as design aids. The ability to detect parameter correlations also seems to be a promising approach for the quality assurance of semiconductor devices.

An important goal of nanoelectronics is the realization of local processing, where data is only exchanged between neighboring elements. But despite the local processing concept, we still want a system to produce global results. Differential equations fulfil this need. Their computational operations are restricted to local elements but their result is strongly dependent on the boundary conditions. An example that also takes self-organization into account is presented next.

The key to local processing and self-organization is given by a quantum-mechanical algorithm: As stated above, the difference between two information vectors can be treated as some kind of system energy, or a potential $V$. It reveals that a self-organization effect can be established by computing a $\Psi$-function from a modified Schrödinger equation, and carrying out an adaption proportional to $\Psi^2$, see Chap. 3. By using this algorithm the input vectors are assigned to global minima. The important advantage of this concept is that not only can local minima be found, and the adaption processes are not very time consuming, but it works locally restricted and fully parallelized. This is an example of using models of solid-state physics also in system-technology concepts.

The structures that are stored inside a self-organized map can be directly used to build nanoelectronic structures, because such networks are topology-preserving systems. To use this effect we have to initialize a homogeneous system with a given starting structure by a learning algorithm. This information can, for example, be stored in a nonvolatile memory device. This method could be extremely important for three-dimensional nanoelectronic systems, which cannot be produced by common lithographic structuring. The topology-conserving feature makes it possible to realize a given topology by using a self-organizing map, if the map already exists as a hardware implementation. The weights of processing elements that contain the structure to be implemented may be set by self-organization. This is how structures may be transferred to a self-organized system that initially was homogeneous. From these, SOMs structures may be transferred to neighboring nanoelectronic systems without using any kind of lithographic process.

## 7.3 Summary

Softcomputing offers interesting opportunities for nanoelectronics, for example fault tolerance, association, self-organization, local adaption, etc. Because these systems are modular, composed of relatively large units, they are also suitable for nanoelectronics. A further advantage is the possibility to build autonomous systems that adapt and program themselves.

# 8

# Complex Integrated Systems and their Properties

Information processing can be divided into three different layers that are reflected by the following representation: The description on the mathematical layer, the data processing with networks that are based on ideal switching devices, data processing with an integrated system that is based on real switching devices. Generally, systems depend on an appropriate architecture, some of them have been described in the preceding chapters.

## 8.1 Nanosystems as Information-Processing Machines

### 8.1.1 Nanosystems as Functional Blocks

Various realizations of microelectronic systems already exist. Most of them can be reduced to the block diagram in Fig. 8.1a. This figure outlines a system that comprises sensitive sensors, effective actuators at the interface to the outside world, low-noise amplifiers, accurate A/D and D/A converters, CPUs, and effective interfaces to the communication networks.

First, the complete task and function specifications of a new system have to be described. The function description has to be realized partially in hardware and pationally software (Fig. 8.1b). The hardware-software codesign is relatively complicated and needs much intuition, since the partitioning is not easily predetermined. The algorithms have to be mapped in a safe and reliable fashion into the system. In the following, a closer look will be taken only at the hardware components. In general, the costs of the hardware components are low in comparison to the overall system. However, the hardware is a key component of the system.

Besides safety and reliability the overall system performance is of general interest. To a first approximation, it can be described as information processing per unit time. Today, high performance is only attainable with integrated circuits, since they are fast, small, reliable, and inexpensive.

124    8 Complex Integrated Systems and their Properties

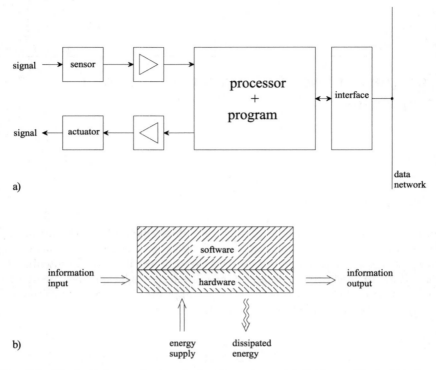

**Fig. 8.1.** Block diagram of microelectronic systems: (a) System with its interfaces and most important functional blocks, (b) System that has been mapped into hardware and software as well as its information and energy flow

High system performance is the objective of system design. Figure 8.1b reveals the information flow from the input to the output. The accompanying information change will be discussed in the next section. Energy is needed for information processing. Thus, there exists an energy flow besides the already-mentioned information flow. The energy has to be supplied as electric energy at the input, while it has to be eliminated as heat at the output.

### 8.1.2 Information Processing as Information Modification

According to the model, the system transforms the information cube to a usually smaller information cube (Fig. 8.2). From the viewpoint of system engineering the following questions arise: What is the minimum of devices needed to process input data into output data? What is the minimum energy needed to gain a higher degree of order and to simplify the data [40]?

The NAND operation (Fig. 8.3a) comprises four input combinations (2 bit) whereas the output has only two output states. The states of the output

## 8.1 Nanosystems as Information-Processing Machines

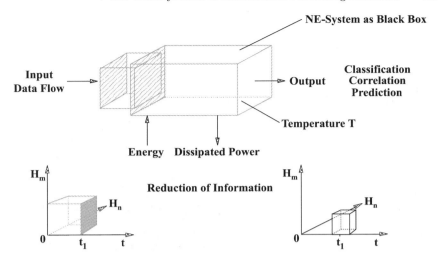

**Fig. 8.2.** Information cube of information processing systems: Degree of parallelism, performance, and complexity

are not uniformly distributed (0.8 bit) and it is impossible to reconstruct the input data from the output data. Thus, information gets lost. In this example information is represented by energy, which has to be dissipated when it comes to information destruction. From another point of view, information loss is another form to tidy up data. Mead and Conway [1] call this from of entropy modification 'logic' entropy. It amounts to the NAND operation to 1.2 bit.

Besides the 'logic' entropy, information processing also has to deal with the so-called 'spatial' entropy. It describes information at its wrong location. In this context, Fig. 8.3b gives an example: A memory with eight memory positions should store two independent bits. This results in 28 possibilities. If the two bits are combined together and form a word the number of possibilities declines to four. This effective combination reduces the 'spatial' entropy by 2.8 bit.

In both cases energy is needed for the information reduction. Today, it is not yet possible to link this information change to technological data in an effective manner. The entropy modification $\Delta H$ accounts for volume changes of the information cube and is generally called negentropy $\Delta H$. According to the thermodynamic model and the information $\Delta H$, the minimum switching energy that is needed to process a single bit amounts to:

$$W = kT ln2 \Delta H. \tag{8.1}$$

This is the minimum energy value. However, higher energy values are needed for safe operation. It has to be assured that faults caused by thermal fluctuations can be compensated by usual fault-correction mechanisms.

# 8 Complex Integrated Systems and their Properties

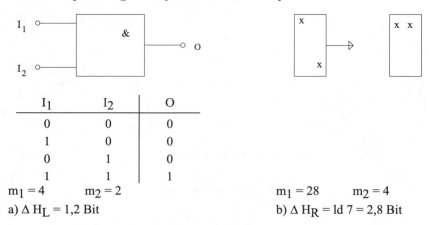

|  $I_1$ | $I_2$ | O |
|---|---|---|
| 0 | 0 | 0 |
| 1 | 0 | 0 |
| 0 | 1 | 0 |
| 1 | 1 | 1 |

$m_1 = 4$   $m_2 = 2$                     $m_1 = 28$   $m_2 = 4$

a) $\Delta H_L = 1{,}2$ Bit                b) $\Delta H_R = \text{ld } 7 = 2{,}8$ Bit

**Fig. 8.3.** Entropy modification in terms of information processing: (a) Logic entropy change of the NAND operation, (b) Spatial entropy change of a memory

Thus, the minimum switching energy $W_S$ amounts to $50\,kT$ as has already been described in Chap. 3.

$$W = 50\,kT ln2 \Delta H. \qquad (8.2)$$

The values of today's microelectronics exceed this value by orders of magnitudes. Nevertheless, concepts are under investigation that do not transform the energy of the lost information into heat. For instance, the conservative Fredkin gate transports the lost information to the outside and eliminates the heat, e.g. via an external resistance or via an antenna.

## 8.2 System Design and its Interfaces

The history of electronics reveals that each technological phase has brought a revolution on the system-engineering level. For instance, large-scale-integration (LSI) came up with the integrated circuits like transistor-transistor-logic (TTL) and emitter-coupled-logic (ECL), whereas microprocessor units appeared with the very-large-scale-integration (VLSI). Therefore, the transition from millions of integrated devices to billions of devices will probably result in a new system-engineering level. Methods for a global performance classification of nanosystems are needed. Their values have to be linked to technological data.

Nanoelectronics has to offer considerably more than the already existing silicon technology in order to be successful. Thus, billions of integrated devices are needed. New design methods have to be developed for the overall design that starts from the single device and ends with complex systems. Technology-invariant interfaces have been introduced to micro- and nanoelectronics.

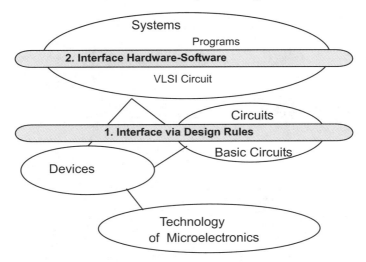

**Fig. 8.4.** Since 1980, the development of microelectronics profits from the separation of system-engineering level and process technology, thanks to Mead and Conway [1]

A considerable milestone of microelectronics was introduced in 1980, when Mead and Conway separated technology from circuit design (Fig. 8.4). In this context, the introduction of design rules is of fundamental relevance: geometric design rules describe the lateral dimensions of the devices. Electric design rules reveal for instance, the resistance and the capacitance of the structures, whereas the technological design rules describe the dimensions and the doping of the structures. The separation of the system-engineering level from the process technology via technology-invariant interfaces has formed the basis for the extraordinary fast development and widespread use of integrated systems.

The compiler is a very important interface between processor hardware and software. This interface has been adopted from computer science to the world of microprocessors. Therefore, the difficult programming of a microsystem via machine code (MC) is obsolete.

What does a nanoelectronic interface look like? The first interface is located between the different technologies and the basic circuitries, similar to microelectronics (Fig. 8.5). A second interface between the basic circuitries and the hardware system is conceivable. Further, a third interface exists between the system and the outside world that concerns the hardware-software codesign (Fig. 8.6). For this task the designer living in the heaven of software has to bridge down to the technologists who are working in the hell of physics. Probably, on the long term, systems will be programmed by their outside world, i.e. systems will be adaptive and autonomous.

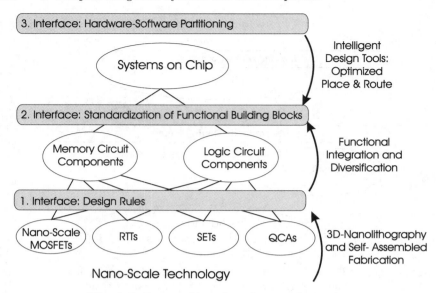

**Fig. 8.5.** Expected interfaces in nanoelectronics

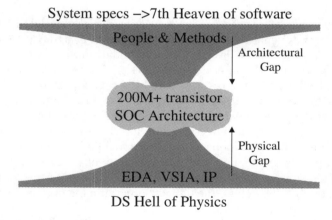

**Fig. 8.6.** In terms of complex systems, hardware and software have to be codesigned. The system demands a reliable hardware base (de Man)

Additionally, technology-invariant system interfaces exist besides the already mentioned design interfaces. For instance, memory applications can be based on different technologies. Thus, a hybrid technology can comprise, on a single substrate, like silicon, different technologies (Fig. 2.15).

## 8.3 Evolutionary Hardware

Another interesting approach investigates hardware concepts that develop with their hardware resources and demands in an autonomous fashion. This is another type of biology-inspired concept. For instance, cells in a nutrient solution form the requested network. This is only possible since each cell comprises the overall structure (see DNA in Fig. 5.3).

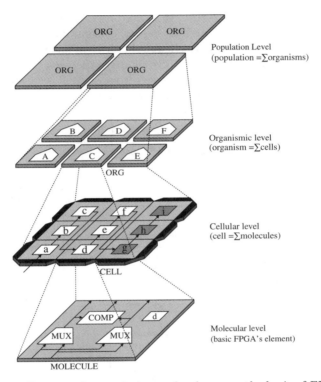

Fig. 8.7. Concept of an evolutionary hardware on the basis of FPGAs

Research focuses on both evolutionary hardware concepts and their realization. One has to differentiate between off-line evolution and on-line evolution. In terms of off-line evolution, the structuring process is simulated on a mainframe computer before it is mapped into the hardware. On the contrary, the on-line evolution takes place on the hardware level. Because of the lack of adequate hardware, up to now, research makes use of field programmable gate arrays (FPGAs) (Fig. 8.7, Pierre Marchall, CSEM [41]). The evolutionary hardware units rely on an appropriate wiring structure and activate adjacent units to solve complex tasks. Once again, the overall structure has to be stored in each individual cell.

## 8.4 Requirements of Nanosystems

Figure 8.8 deals with architectures of some nanoelectronic systems and describes their implementations. From today's point of view the following applications might be realized in future: Demanding process control, strategic modeling, autonomous systems, self-assembling systems. This means more intelligence must be integrated into the systems via increasing numbers of devices. This is challenging for nanoelectronics.

In terms of system architecture, the data-processing performance, the degree of parallel data processing, the power consumption, and wiring constraints are of major concern. In particular, in terms of image-processing applications, interesting approaches with distributed processor structures already exist. Another promising approach comes from the fields of softcomputing, since artifical neuronal networks lead to new concepts.

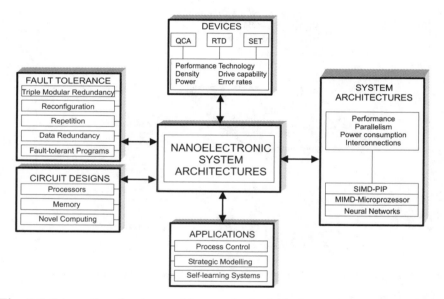

**Fig. 8.8.** Integration of system architecture in the field of nanoelectronics, according to T. Fountain (UCL)

Fault tolerance is an important issue when it comes to systems that comprise such a huge number of devices. In this context, the following concepts already exist: Modular redundancy, free reconfiguration, recurrence, data redundancy, and fault-tolerant programming. In particular, the last-mentioned concept is of fundamental relevance for today's microcomputers, but it would be inadequate for nanoelectronics. The concrete hardware realization needs new concepts on the processor and memory level as well as a new data-processing fashion that might be based,for example on fuzzy sets.

The system-engineering level has to deal with the challenging implementation of the huge number of devices. The high complexity has to be cut down in some kind of functional blocks. From the architectural point of view, regular and modular arrangements are advantageous. Obviously, the high packing densities require a very low power dissipation. Local processing and sparse interaction is attractive, since wiring is a critical issue in relation to the high complexity. A global wiring method is almost not realistic. The similar statements are true for the linkage to the outside world: The number of pins to the outside is limited as well. Thus, an autonomous system should be ideal. From the already-mentioned system properties, further natural demands arise: The system should be self-testable and fault tolerant. Additionally, the system should be adaptive, self-optimizing in order to autoadapt itself to the environment. In a more general term one can state: Local processing should conclude in a global effort.

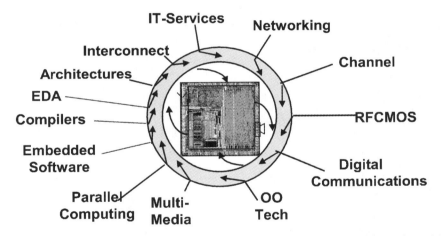

**Fig. 8.9.** The design of today's complex systems comprises a large number of special fields, e.g. the SoC of a mobile communication system, according to de Man [42]

However, the design of complex systems comprises many areas as Fig. 8.9 shows for the example of communication systems on a chip. This challenge has to be met during education or on-the-job training.

The system concepts of nanoelectronics are at their early stage, they are still under investigation. Therefore, it is impossible to deliver complete architectures, merely visions.

## 8.5 Summary

Information processing of nanosystems can be described as a modification of the information cube. The accompanying information loss contributes to the total power dissipation of the system. However, the major part of the dissipated energy results from the power supply of the switching devices.

Technology-invariant interfaces will be of fundamental relevance for the design of nanosystems. In contrast to microelectronics, the interfaces will be shifted in the context of the engineering level towards higher complexities. Thus, technology-invariant interfaces have to be introduced between the functional blocks and the overall system in order to integrate different kinds of technologies.

# 9

# Integrated Switches and Basic Circuits

Up to now we considered information processing as an abstract procedure. However, the objective is to implement the concept of an information-processing system into real chips. For the implementation we need an appropriate technology that should be right for mass production.

Technical information processing is based on real physical structures: Information-processing machines primarily comprise switches and wiring structures. In this context we will show some examples of the classic semiconductor technology. Additionally, we will discuss the differences between an ideal and a real switch, as well as ideal and real wiring.

**Fig. 9.1.** Abacus with discrete atoms, IBM [43]

## 9.1 Switches and Wiring

### 9.1.1 Ideal and Real Switches

What characteristics does an ideal switch have? In terms of the abacus it might be a tag that moves without friction on its bar, Fig. 9.1. In terms of today's computers an ideal switch has an infinitely high resistance in its open state and an infinitely low resistance in its closed state (Fig. 9.2a).

Nevertheless, a real switch can not provide by any means this high resistance ratio: In practice the two binary states are approximated by relatively high and relatively low resistances. Moreover, these resistances are not constant, since they scatter around mean values. In Fig. 9.2b the shaded area reveals this scattering. Furthermore, a real switch consumes during each switching event the energy $W_S$. Only after consuming $W_S$ does the switch react with a delay of $t_d$ (Fig. 9.2c).

In terms of thermodynamics the minimum energy for the processing of 1 bit derives from the above considerations in connection with (8.1) and $S = W/T$:

$$W = 2kT\ln 2. \tag{9.1}$$

Even so, this value is not an absolute limit. For instance, the quantum computer can evaluate data without any loss of energy. In general, it is im-

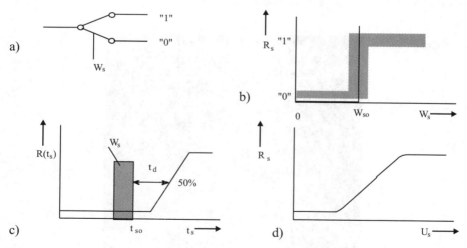

**Fig. 9.2.** Properties of an ideal and a real switch: (a) ideal switch with two states, (b) real switch with finite resistance that scatters around its mean value, each switching event consumes the energy $W_S$ that also scatters around its mean value, (c) the real switch reacts with a delay of $t_d$ after consuming the switching energy $W_S$, (d) the current/voltage characteristics of a real switch might be linear to some extent so it can be exploited for amplification purposes

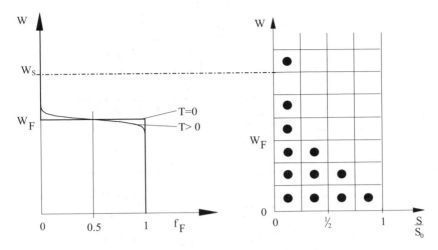

**Fig. 9.3.** Switching energy $W_S$ in the Fermi distribution and in relation to the particles' energy. On the right-hand side the single particle above $W_S$ could flip a switch because of its thermal energy

portant to know that each switching event in the classical sense or on the macroscopic level depends on a certain amount of energy.

The interference of thermal fluctuations with the switching processes accounts for the energy distribution of the particles: It depends on the probability of particles that have a higher energy than the switching energy $W_S$ (Fig. 9.3). It follows from the (3.30):

$$P = e^{-\frac{W_S}{kT}}. \tag{9.2}$$

The value P in (9.2) is called the Boltzmann factor. Even for a very high value of $W_S$ the probability for errors does not decline to zero, since at all times very few particles might have sufficient energy to overcome $W_S$ which may cause errors. As will be derived in Chap. 15 technical systems use switching energies of at least $50\,kT$ to be on the safe side.

This amount of switching energy does not automatically contribute to the dissipated electric power during switching. One half of a capacitor's energy is lost by its ohmic resistance during an instant discharge. This is also true for the charging of a capacitor,

$$W_V = \frac{1}{2}CV^2. \tag{9.3}$$

On the other hand, if the capacitor is charged in a slow way with many intermediate steps $\Delta V$ the dissipated power reduces to:

136    9 Integrated Switches and Basic Circuits

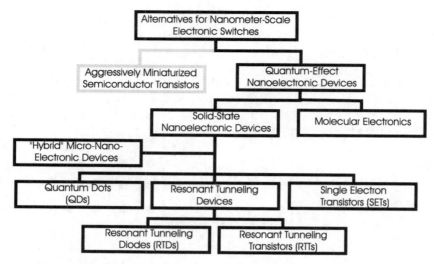

**Fig. 9.4.** Variety of nanoelectronic switching elements, M. Forshaw

$$W_V = \frac{V}{\Delta V}\frac{1}{2}C\Delta V^2 = \frac{1}{2}CV\Delta V. \tag{9.4}$$

This specific switching process is called adiabatic switching and may be relevant to low-power circuits [44].

The restrictions of a real switch cause several disadvantages within electric circuitries. The drift of resistance values should not cause any error, which is a critical issue especially for analog circuits. The delay times should not interfere with the signal-processing sequences: As a rule, the correct timing is ensured via external clocking.

The characteristic values of switches and gates, namely the delay time $t_d$ and the dissipated power $P_V$ evolve from the behavior of the real switch. The dissipated power refers to the power that is consumed during the switching event. Both values form the power-delay diagram and should be kept as small as possible.

According to Keyes the output of a switching element has to be isolated from its input in a suitable manner [45]. The three-terminal transistor complies this constrain whereas the two-terminal tunneling diode does not. Therefore, up to now two-terminal devices remain relatively insignificant.

Figure 9.4 summarizes the landscape of nanoelectronic-oriented technologies. The following chapters will discuss them in detail. A steady movement towards smaller structures keeps device shrinking alive. Besides the high packing density future technologies have to fulfil several criteria: Circuit design should be easy to handle, the yield should be very high, the supply voltage should be below 1 V, the circuits should have a high noise immunity, the quiescent power consumption should be practically zero, and the technology should be suitable for mass production.

However, microminiaturization meets its limits within the classic switching devices: Therefore new principles that are based on quantum effects have to be developed for future nanometric switching devices. These devices will not only have a high packing density, but probably will also show better electrical properties. From today's point of view, molecular electronics that reaches up to the physical limits will become feasible.

Dealing with such large numbers of switching devices as in nanoelectronics, one has to expect defective devices within the circuitry. The only solution for this problem is that redundant devices will have to replace them. At this solution, additional switching devices are not problematic, but the complex wiring for the redundant part will be.

## 9.1.2 Ideal and Real Wiring

System engineering reveals that not only advanced switching devices are needed, but also effective wiring structures. The properties of ideal wiring can be easily defined: First, it should be feasible to connect all nodes with each other. Secondly, the transmission properties should be ideal, i.e. no resistances, capacitances, inductances, so that the speed of light defines the time delay of all connections.

Today's microelectronics exclusively makes use of two-dimensional circuitries that are based on multilayer wiring, in order to save area and to keep timing delays short. Recent technologies have to deal with high current densities that tend to cause electromigration, which can damage the wiring. Copper wires instead of aluminum wires can overcome this problem. Cryogenic conduction can reduce the wiring resistance to approximately zero, but comes with several other problems. Capacitances can not be omitted as well as inductances. However, in many cases inductances can be neglected. Both items define the characteristic wave impedance that is in the order of a few tens of ohms. The characteristic wave impedance itself determines the lower signal levels.

Several approaches have already been attempted for real three-dimensional circuit integration. Intelligent sensors are based on multilayer designs with integrated transistors and complex wiring structures. Other approaches build stacks of single chips. In this case, the interconnection facilities are limited, since wires can only be attached from the outside.

In the fields of nanoelectronics some new approaches already exist. M. Forshaw (UCL) for example, suggests thin silicon chips that might be interconnected with conducting polymers. The silicon substrates have to have perpendicular conducting channels so that they can be interconnected through several layers. The schematic view of Fig. 9.5, for example, presents a concept for an electronic eye.

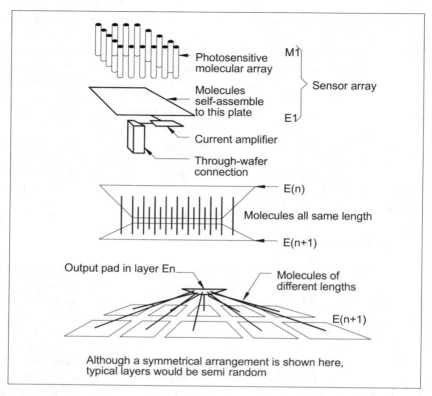

**Fig. 9.5.** Layer structure of the electronic eye. The photosensitive sensors and the interconnects are realized with molecular electronics, whereas the circuit itself is based on silicon, proposal of M. Forshaw, UCL

An electronic eye implements the most important features of an eye into an integrated system. Because of its complexity it is a challenge for nanoelectronics, and it is an important subsystem for robotics and traffic engineering. Today, electronic eyes are implemented as a two-dimensional system in silicon: The signal preprocessing unit is integrated beneath the photodiodes and all signals have to leave the circuitry in a lateral manner. This is less problematic for three-dimensional arrangements.

The arrangement in Fig. 9.5 is of general interest for information-processing purposes. Two-dimensional information is fed into a processing layer structure. On the long term it seems to be feasible for neuronal networks to change to three-dimensional architectures that would offer the opportunity of three-dimensional signal processing. In this context, interconnects should be dynamically configurable (Chap. 4).

Another interesting network approach firstly marks the points that should be connected with DNA molecules. Then, self-organizing DNA strands set up

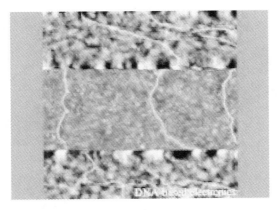

**Fig. 9.6.** DNA molecules for wiring purposes, Storm, de Vries, Dekker, TU Delft, NL

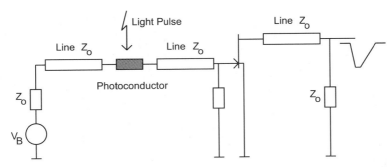

**Fig. 9.7.** Clock distribution via light: Light pulses control photoresistors, R.A. Soref

the interconnections between the marks. The DNA network might be coated with metal in order to enhance the electric conductance (Fig. 9.6). Other promising wiring concepts are based on light beams [46]. Below, two examples will be given.

Beside these two relatively complex examples, a straightforward solution mounts a laser diode into the chip housing. Light flashes control photoresistors that discharge integrated wires (Fig. 9.7). The charge itself controls transistors and results in electric pulses that clock the system.

Figure 9.8a shows a reconfigurable optical network. Depending on the applied electric fields, the input signals appear at the corresponding outputs. An optical connection network between two silicon chips can be realized with a reflection hologram. The input consists of integrated laser diodes that expose light on a reflection hologram. From there, the light is projected onto the pho-

140    9 Integrated Switches and Basic Circuits

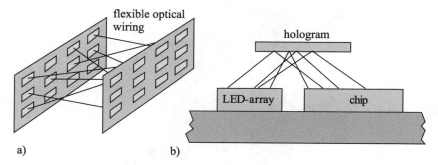

**Fig. 9.8.** Optical flexible wiring: (a) controllable lens system, (b) hologram

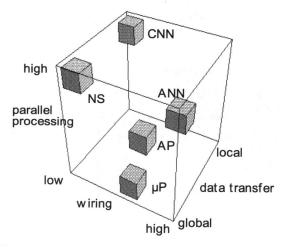

**Fig. 9.9.** Classification of different concepts in terms of parallelism, wiring and data transfer, $\mu P$: microprocessor, AP: array processors, ANN: artificial network, CNN: cellular neuronal network, NS: objective of nanoelectronic systems

todiodes of the second chips (Fig. 9.8b). This leads to a multiple connection network between both chips.

The importance of wiring has already been discussed in Chap. 7 in connection with the properties of complex systems. Figure 9.9 reveals the different system concepts in terms of parallelism, wiring, and data transfer. The microprocessor is included in the three-dimensional diagram as a benchmark. The wiring complexity is relatively high for both array processors (AP) and artifical neuronal networks (ANN). It is relatively simple for cellular neuronal networks. They only serve for simple and mostly very specific tasks. The objective of nanoelectronic systems is to establish high parallelism and global

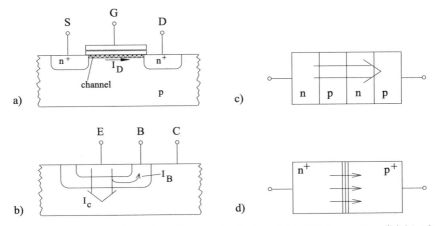

**Fig. 9.10.** Examples of classic solid-state switches: (a) MOS transistor, (b) bipolar transistor, (c) thyristor, (d) tunneling diode

data transfer on the basis of a local wiring network. These constraints have to be met by future system concepts that are still under investigation.

According to M. Forshaw the following theses can be claimed for the redundancy of nanosystems: A compact two-dimensional arrangement can only include a few redundant cells, which results in an illdefined nanoelectronic system. For certain, a widespread arrangement can keep more redundant cells, but will be ineffective because of the widespread wiring. The only remaining solution for an adequate high-performance system is a three-dimensional integration that offers convenient wiring conditions for all cells, especially for the redundant ones. Nevertheless, up to now only two-dimensional CAD tools exist so that three-dimensional tools still have to be developed and they will be quite expensive.

## 9.2 Classic Integrated Switches and their Basic Circuits

### 9.2.1 Example of a Classic Switch: The Transistor

The main purpose of electronics is to control the current flow. Electromechanical switches like relays formed the starting point and have not been replaced in all areas by solid-state switches, because of their enhanced electrical properties. Microsystems are based on solid-state switches so that mechanical switches have nearly no significance for them. For instants, Fig. 9.10 reveals the cross sections of four classic solid-state switches [47].

The MOS transistor consists of a p-doped region and a metallic gate that are separated by a dielectric material (Fig. 9.10a). A positive voltage causes, in connection with the field effect, a conducting electron channel at the surface of the semiconductor. The conducting channel connects both n-doped regions of the transistor in the electrical sense. Without an applied gate voltage both regions are insulated from each other. Thus, the gate voltage opens and closes the switch, the switching process is at the surface of the chip.

Figure 9.10b depicts the schematic view of a bipolar transistor. It consists of three different doped areas. The base current $I_B$ flows in the forward direction of the emitter-base diode. The relatively high emitter current does not leave the transistor at the base side, but flows mainly to the collector side, since only very few carriers recombine in the thin base. The emitter injects carriers to the collector. Without a base current there is no collector current. Therefore, the bipolar transistor is a switching device controlled by a current. Its switching process takes place in the volume of the chip.

A four-layer npnp-structure forms the thyristor switch (Fig. 9.10c). A small current between the inner layers puts the thyristor into the conducting state. Thyristors belong to the group of power-electronic devices, however, microelectronics also makes use of them: Chap. 2 describes how thyristors regenerate light signals at fiber optics. Light beams can switch thyristors that, for instance, drive LEDs. The thyristor is advantageous since it connects fully through and delivers the complete output signal. This layer structure also appears as a parasitic device within CMOS technology. The so-called latch-up emerges between a well and its neighboring transistors and can destroy the chip due to the high current by a short-circuit.

A highly doped pn-junction makes a tunneling diode. In general, it is called an Esaki diode (Fig. 9.10d and Fig. 12.7). The applied voltage controls the current of the tunneling diode. The control depends on two effects: First, the electric field deforms the potential barrier, which reduces its effective thickness. Secondly, the band edges move towards each other due to the applied voltage. The carries can only tunnel if the other barrier side offers a free band. Obviously, the position of bands has a strong impact on the current through the diode. Tunneling devices will probably have a strong impact on nanoelectronics, since the tunneling takes place very rapidly and without any loss of energy.

Since the tunneling diode is a two-terminal device the enhanced switching properties suffer from the fact that no input-output insulation is feasible. New device structures avoid this drawback, for example, by combining two serially connected tunneling diodes. A control electrode takes care of the inner potential voltage of the center. In this structure the resonant tunneling transistor is a promising candidate.

### 9.2.2 Conventional Basic Circuits

Circuit-design techniques for binary logic circuits will be taken for granted and only their basic principles will be reviewed in the following. The first semiconductor circuits based on logic circuits had been already used by Zuse for his relay-logic (Fig. 9.11a). Instead of electromechanical switches, transistors like the bipolar transistor (Fig. 9.11b) and field-effect transistor (Fig. 9.11c) had been used initially [48].

In some ways the transistor has poor properties in comparison to a mechanical switch. For instance, gates have to deal with the problem of residual voltage: The active transistor causes a current that results in a voltage drop across the transistor. This voltage drop has to be smaller than the threshold voltage of the following transistor stage, otherwise the subsequent stage would always be activated [49].

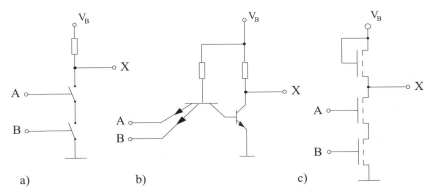

**Fig. 9.11.** Different NAND gate realizations: (a) electromechanical switch (relays), (b) bipolar transistor, (c) MOS transistor

A simple response to this problem could be a relatively high load resistance $R_L$. But this solution is disadvantageous for the charging time of the output capacitance $C_L$ that gets charged via $R_L$. Besides the problematic residual voltage, one has to deal with the high power dissipation.

Two possible solutions of this dilemma might be the following: The first solution passes from the static to the dynamic circuit design technique. Thus, the output capacitance is only (dis-)charged if the transistor in the drive or load branch is activated (Fig. 9.12a). The charge on the capacitance $C_L$ represents the logic state of the gate. Due to charge leakage, such dynamic circuits depend on a minimum clock frequency so that the charges get regenerated before it completely disappears.

The second solution is based on the relatively complex CMOS technology with its complementary transistors (Fig. 9.12b). The quiescent current of this circuit technique can be considered to be zero for all practical purposes. Both the charging and discharging of the capacitance $C_L$ is done by low-resistance switches. Due to these outstanding properties, this technique is presently the most popular one.

Another alternative is based on transfer of electric charge. In this context the field effect transistor is a very suitable device, due to its symmetric be-

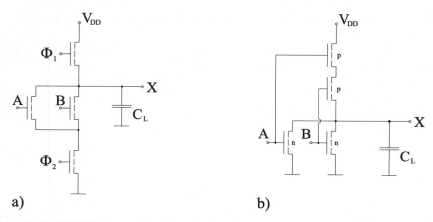

**Fig. 9.12.** NAND gate with relatively low static power consumption and without residual voltage: (a) dynamic gate, (b) CMOS gate

havior (Fig. 9.13). Both the drain and the source contact can serve as a signal input. Together with an appropriate clock supply, relatively complex gates can be realized with a reduced number of devices (Fig. 9.13b). A special type of this circuitry are the so-called switched capacitor filters.

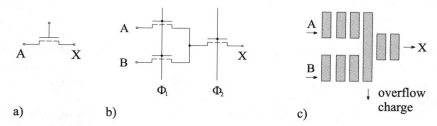

**Fig. 9.13.** Controlled logic: (a) MOS transistor as switch and transfer element, (b) OR gate composed of transfer elements, (c) OR gate on the basis of charge coupled devices (CCD)

A more advanced approach gives up the transistor structures and utilizes charge coupled devices (CCDs). Within the CCD structures charge is moved, brought together, and generated at the surface of the substrate (Fig. 9.13c).

Up to now, preferred design techniques are based on voltage-mode circuitry. However current-mode circuitries are also an interesting approach that might be advantageous for future low-voltage applications.

### 9.2.3 Threshold Gates

Threshold gates are promising concepts that are probably suitable for nanoelectronics [50]. Threshold logic and its application possibilities had already been investigated before. It turned out that binary techniques are more effective. However, this strategy might change when it comes to nanoelectronics.

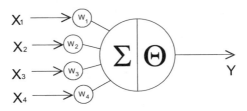

**Fig. 9.14.** Threshold gate with four inputs

Threshold gates can evaluate any Boolean operation. For instance, Fig. 9.15 shows the threshold values for a logic OR gate and a logic AND gate, respectively. With respect to the threshold gate in Fig. 9.14 these two logic gates have only two inputs,

$$y(\chi) = \text{sign}\ (\chi - \Theta) = \begin{cases} 1 \text{ if } \chi \geq \Theta \\ 0 \text{ if } \chi < \Theta \end{cases} \quad (9.5)$$

$$\chi = \sum_{k=1}^{N} w_k \cdot x_k, \qquad x_k = \{0, 1\}$$

$$w_k = \{0, \pm 1, \ldots, \pm w_{max}\}$$

$$\Theta = \{0, \pm 1, \ldots, \pm \Theta_{max}\}\ .$$

Threshold gates are very suitable for full-adder designs (Fig. 9.16), due to their reduced logic depth and high parallelism. The circuit in Fig. 9.17 makes extensive use of these advantages. In this context, Chap. 12 reveals an elegant realization.

The bigger the adder, the more threshold values are necessary. The range of values of the above example goes from 1 to 9. Since it is difficult to define these threshold values within today's technologies, they do not have any significance. In terms of nanoelectronics the number of threshold values automatically corresponds to the number of switching devices, which solves the problem of the lack of precision. The longing for higher clock speeds and shorter cycle times in connection with higher integrated CPUs require robust gates that evaluate their operations within a few steps. All these claims can be met by threshold gates, in spite of their known drawbacks.

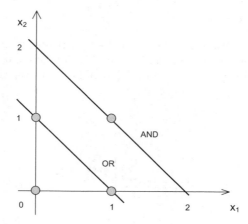

**Fig. 9.15.** Threshold values: threshold $\Theta \geq 1$: OR gate, threshold $\Theta \geq 2$: AND gate

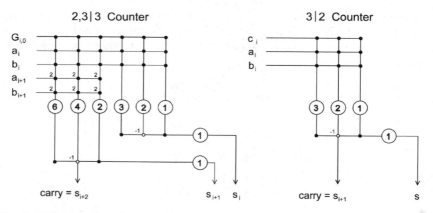

**Fig. 9.16.** Examples of a one-bit full adder: Both solutions have a logic depth of two. The range of threshold values increases with the logic width of the adder

In contrast to systolic system configurations, systolic arrangements on the bit level primarily perform simple operations. Usually, one cell receives several one-bit inputs from its neighbor cells. This information is processed on a low level. The critical path of a single cell is in the boundaries of two to three logic gates. On the edge of each cell several D-latches are responsible for synchronization purposes. This concept omits long wiring distances and is usually applied to pipelined adders and multipliers.

Figure 9.17a shows two systolic 8-bit ripple carry adders. Input data and the sum bits are propagated in the vertical direction, whereas the carry bits propagate in a diagonal fashion. The adder in Fig. 9.17b reveals how the delay

## 9.2 Classic Integrated Switches and their Basic Circuits

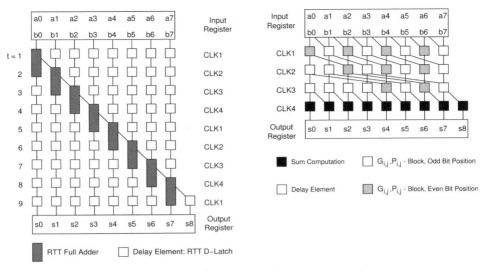

**Fig. 9.17.** Parallel adders with reduced logic depth

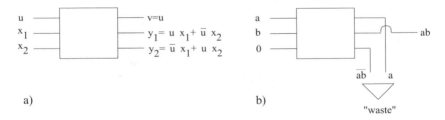

**Fig. 9.18.** Principle of the Fredkin gate: (a) Example of the Fredtkin gate, (b) realization

time can be reduced by parallelizing the sum operation of two bits. Threshold gates proved to be suitable for these operations.

### 9.2.4 Fredkin Gate

Obviously, a conventional AND gate suffers from the information loss. It is not possible to reconstruct the input signals from the output signal. According to Chap. 8 (Fig. 8.3) the loss of information accompanies power dissipation. Therefore, the questions arise if it is feasible to build logic gates that do not suffer from information loss and if data processing always accompanies power dissipation. Actually, there is a gate family that processes information without the loss of energy: The so-called Fredkin gate.

The Fredkin gate interlinks the input data that appear at the output in varied manners (Fig. 9.18a). The basic idea is to forward only the information that is needed in a following stage. The unused information is collected as 'waste' at a bus line that also serves as a terminal resistance. Therefore the Fredtkin logic remains, in its inside, cool. Within the circuit no power dissipation appears in regard to information loss. Figure 9.18b reveals, for instance, an AND operation.

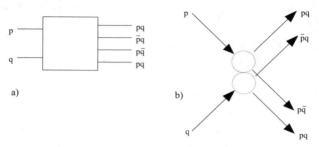

**Fig. 9.19.** Interaction gate: (a) schematic view, (b) Realization with balls on the basis of the elastic collision. According to Drexler, such arrangements are feasible in terms of nanomechanics

Figure 9.19 illustrates an interaction gate on the basis of a Fredkin gate with balls. The elastic collision does not cause any power dissipation. The information garbage has to be collected and led to an output. This form of information processing is reversible.

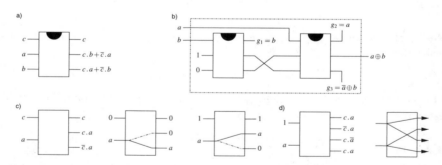

**Fig. 9.20.** Typical Fredkin gate configurations: (a) A binary conservative logic gate, (b) A XOR gate from Fredkin gates with the garbage lines g1/g3 and the data lines 0 and 1, (c) A unary conservative logic gate, (d) interaction gate

At first sight, this concept seems to have only theoretical importance. However, it turned out to be very suitable for molecular- and nanoelectronics. The Fredkin gate logic prevents the heating of molecules during information processing. This point is an important issue, since it is very difficult to cool molecular structures. Information waste is led to the edge of the gate and cannot heat the molecule. In terms of polymer electronics, electrons do not emit their energy to the molecule or its surrounding, but take it to the outside to the waste-bus.

In general, it is challenging to investigate new concepts in electronics that exploit these principles that might lead to advantageous circuitries.

## 9.3 Summary

An excellent integration technique depends on both good switching devices and good interconnection structures. For many decades the same principles of solid-state switches, like the field effect transistor and bipolar transistor have been improved. However, in recent years new circuitry concepts have not been put into practice. Therefore threshold logic lives in the shadows. The switches themselves are not important for the application, but their basic circuitries are. As a rule, a better power-delay product results in higher hardware expenses.

# 10

# Quantum Electronics

Traditionally, microelectronics is based on scaling towards smaller structures. When the dimensions of these structures reach the nanometric scale, microelectronics has already transformed to nanoelectronics. Future systems that are composed of these nanometric structures require both novel architectures and modern switching devices [51]. The behavior of these new devices can primarily only be explained with quantum-mechanical models. The second half of this chapter deals with some important nanoelectronic switching devices.

## 10.1 Quantum Electronic Devices (QED)

### 10.1.1 Upcoming Electronic Devices

In the classical sense, a semiconductor switch depends on electrons that can be characterized by their elementary charge q. This simplified model is becoming

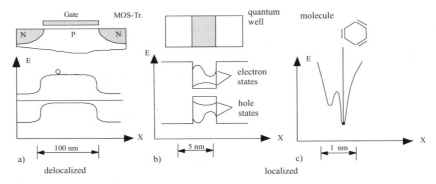

**Fig. 10.1.** Charge storage: (a) homostructure, (b) heterostructure and (c) molecule (simplified). The switching event of the MOS transistor is characterized by many electrons, whereas other devices only change the state of a single electron

technically obsolete when it comes to nanoelectronics. Nanoelectronics has a much higher packing density than conventional FET electronics, thus it is most important to work towards new architecture concepts. At present, the minimum feature size of integrated circuits is ruled by the delocalized charge storage. Its relatively high extension depends, amongst other things, on the Debye Length (1.2) and is in the range of almost 100 nm (Fig. 10.1). However, within the boundaries of classical physics local charge storage as is known, e.g. from molecules, only requires few nanometers.

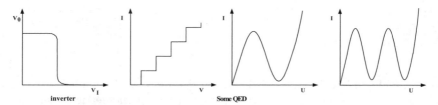

**Fig. 10.2.** Some I-V characteristics of quantum electronic devices (QED)

These dimensions can be reached in terms of very thin semiconductor and insulator layers. Devices that have at least one dimension in the order of magnitude of the electron wavelength are called *mesoscopic elements*. One of these elements is the quantum well with its characteristic thin semiconductor layer. Another example is the tunneling structure with its typical narrow insulator layer. With respect to quantum-mechanical devices, the behavior of electrons in mesoscopic structures will be analyzed in the following.

Both mesoscopic and quantum-effect devices are featured with enhanced properties. The single-electron-transistor (SET) emerges from the MOS transistor if it is scaled down and the two pn-junctions are replaced by tunneling elements (TE). Within the SET the charge of one electron controls the current of single electrons. In a similar way flux-quantum devices control the number of magnetic flux quanta in a ring structure via two tunneling elements. The dimensions of such a configuration need not be in the nanometric range, which may be advantageous. The focus of the next chapters lies on these technologies.

A quite new approach is based on regular quantum-dot structures in which single electrons are moved locally. In the literature they are described as quantum cellular automata (QCA). Data is not propagated in the form of charge along metallic wires, but transformed by information patterns. The dimensions of the quantum dots are in the domain of a few nanometers, which results in very high packing densities and leads to a real nanoelectronic concept. Such a concept can also be realized with magnetic flux quanta, which is of fundamental interest. Although its realization is in the distant future, we will take a closer look at it in the following.

## 10.1 Quantum Electronic Devices (QED)

Quantum-effect devices show enhanced I-V characteristics. The typical inverter characteristics of the MOS technology can also be found within quantum-dot devices. In many cases the current of quantum-effect devices is a staircase function of the voltage. The grid of the stair function depends directly on physical quantities (Fig. 10.2).

As a general rule, devices with tunneling elements (TE) has I-V characteristics with a negative differential resistance (NDR) segment. This regime can be used for information storage and processing purposes. Several NDR regimes can be achieved with appropriate band structures that might be used for functional integrated gates.

The same circuit architectures that are typically used for conventional circuit designs can also be utilized for circuit designs that are based on quantum-effect devices. However, more appropriate architectures that directly benefit from the enhanced device characteristics have to be taken into account, e.g. for multivalued logic applications.

Nevertheless, an interface has to be established between any new architecture and the conventional Boolean algebra. In this connection, universal gates that can be programmed by a control signal are advantageous. A comprehensive example is the majority gate. Its output depends on the majority value of all input values. It is an interesting gate for fault-tolerant systems. Additionally, it can also be considered as a universal gate: If the control signal is connected to the logic low level its logic function is AND. On the contrary, if the control signal is connected to the logic high level, its logic function is OR (Fig. 10.3).

Other approaches take advantage of the material wavefunction of the electron, which also leads to new switching elements. Below, two of these approaches are discussed in detail.

### 10.1.2 Electrons in Mesoscopic Structures

The band diagram of a semiconductor depends on its crystal structure and on the material wavefunction of the electrons [47]. The band diagram reveals the

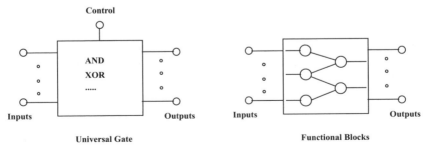

**Fig. 10.3.** Universal gate and a complex functional block

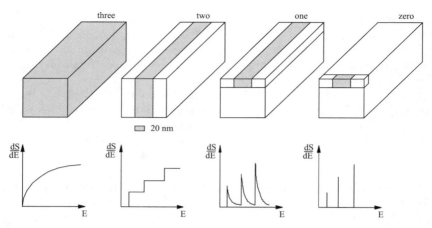

**Fig. 10.4.** Quantum structures with different dimensions: normal solid-state body, quantum well, quantum wire, and quantum dot, additionally their density of states are illustrated. The grid of the steps depends on the size of the quantum structure

energy levels that can be occupied. The density of states of these levels depend on the size of the crystal. An extensive three-dimensional body produces a three-dimensional k-space. The energy $E$ of the electrons can be derived from solid-states physics as:

$$E = \frac{\hbar^2}{2m}(k_x^2 + k_y^2 + k_z^2) = \frac{\hbar^2}{2m}|\mathbf{k}|^2 . \qquad (10.1)$$

With respect to the three-dimensional k-space the possible states can be expressed as:

$$S = \frac{V_T}{\pi^2}\frac{1}{3}\left(\frac{2mE}{\hbar^2}\right)^{3/2} . \qquad (10.2)$$

The density of states follows as:

$$\frac{dS}{dE} = \sqrt{2}\frac{V_T}{\pi^2}\frac{m^{3/2}}{\hbar^3}\sqrt{E}. \qquad (10.3)$$

The density of states is proportional to the square root of the energy of the electrons. This function describes the well-known Boltzmann relation of classical physics (Fig. 10.4).

A three-dimensional relatively large ashlar (3D potential well) has a steady $\sqrt{E}$ characteristic for the density of states, whereas the density of states characteristics becomes discontinuous if at least a single dimension is scaled down to the domain of the material wavelength of an electron. This leads to quantum layers (2D potential well) and appears, e.g. in the inversion layer of a

MOS transistor. Further steps lead to the quantum wire (1D potential well) and quantum dot (0D potential well).

An electron space that is limited in one direction leaves an unlimited two-dimensional k-space for the electron. In this case of a quantum layer the energy of the particle is:

$$E = \frac{\hbar^2}{2m}\left(k_x^2 + k_y^2\right) + E_i. \tag{10.4}$$

Along the reduced dimension only states that match with the wavelength of the electrons can appear. Therefore, discrete energy levels $E_i$ appear. Because of the two-dimensional k-space the density of states $S$ is directly proportional to $E$. If this expression is derived to $E$, $dS/dE$ is independent of the energy (Fig. 10.4).

A further step leads to the so-called quantum wire, which leaves only one dimension of freedom to the electron. Consequently, the energy can be expressed as:

$$E = \frac{\hbar^2}{2m}\left(k_x^2\right) + E_{in}. \tag{10.5}$$

Under these conditions the density of states is for the one-dimensional k-space inversely proportional to the square root of the energy.

A restriction of all three dimensions of the electron results in the so-called quantum dot. Under these conditions only discrete energy levels can appear:

$$E = E_{imm}. \tag{10.6}$$

In this case the diagram of the density of states shows only discrete lines, similar to a hydrogen atom or a molecule.

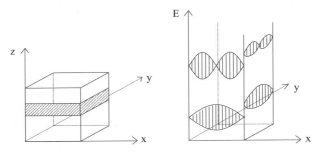

**Fig. 10.5.** Quantum dot and some of its states in the E(x,y) characteristics

Such a quantum dot can be realized from small semiconductor columns (Fig. 10.5). In general, the lateral dimensions are adjusted so that the energy value of an electron is about one electron volt. The third dimension depends on a very thin epitaxial layer. With respect to this dimension the electrons own at least twenty electron volts. This results in a clearly arranged energy spectrum since vertical energy levels only appear at higher energy levels. These arrays can be tuned to desired wavelengths and emit sufficient power for laser applications.

In this connection the relatively abrupt junction of the quantum well is of fundamental relevance. In general, the abrupt junction depends on specific doping profiles of the semiconductor. They have to be essentially smaller than the material wavelength of the electrons. Such abrupt junctions can be realized in $GaAs$ via molecular beam epitaxy. Within the silicon domain it is very difficult to realize these abrupt junctions. Additionally, such silicon-based devices require relatively low temperatures so that at present $GaAs$ has some advantages over silicon with respect to nanoelectronics.

## 10.2 Examples of Quantum Electronic Devices

### 10.2.1 Short-Channel MOS Transistor

Since the conducting channel of a MOS transistor is very thin, its inversion layer can be considered as a two-dimensional electron gas in a quantum layer. Therefore different stationary electron waves can be traced in its cross section (Fig. 10.6).

The discrete energy levels of such a quantum layer appear at low temperatures. Experimental results reveal oscillations in the I-V characteristic of short-channel MOS transistors (Fig. 10.7).

These oscillations are due to quantum effects. However, careful research has revealed that an exact interpretation of these effects is relatively difficult. They can probably be explained by the Coulomb blockade, however, some signs are also emerging that quantum chaos may model these effects.

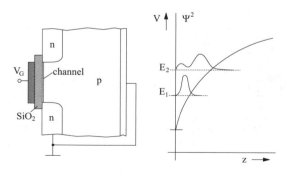

**Fig. 10.6.** Appropriate doping and adequate dimensions transform the inversion layer of the channel to a quantum layer with discrete energy levels

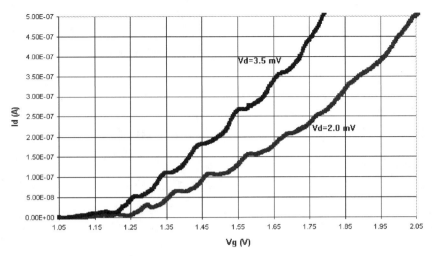

**Fig. 10.7.** I-V characteristic of a MOS transistor with a channel length of 30 nm: Quantum effects appear in terms of oscillations, G. Wirth [52].

### 10.2.2 Split-Gate Transistor

The split-gate transistor is an interesting example of mesoscopic systems (Fig. 10.8). The small dimensions of the gate slit result in different wave configurations that have a direct impact on the conductance between the source and drain contacts. The gate voltage alters the shape of the potential well. Mainly, the depth of the potential well is modified so that more eigenstates can be occupied. Consequently, the conductivity between S and D increases. This increase does not take place in a continuous way, but is a staircase function of the gate voltage. It is of fundamental interest that the step of the staircase function only depends on physical quantities, namely the elementary charge q and the elementary quantum of action h. Therefore, parameter fluctuations that, in general, are problems for sub-$\mu m$ circuit designers are not an issue for the split-gate transistor.

The grid of the staircase function can be derived heuristically from the uncertainty principle:

$$\Delta E \Delta t \geq h. \qquad (10.7)$$

The energy is proportional both to the voltage V and to the charge 2q, because each state below the Fermi level can be occupied by two electrons. The time t depends on the transport time of an electron and can be expressed as $q/I$.

With respect to the uncertainty principle, the conductance can be expressed as:

$$\Delta G = \frac{I}{U} = \frac{2q^2}{h}, \tag{10.8}$$

which is equivalent to the step of the staircase function.

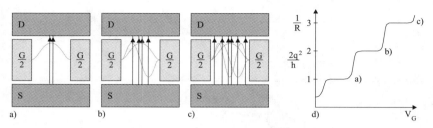

**Fig. 10.8.** Mesoscopic element: Split-gate transistor with three different operation conditions (a-c) and its $GV_g$ characteristics (d) that can be described by Schrödinger's equation

A similar behavior shows the von Klitzing structure that comprises a MOS transistor with four connections at the channel area. Within a magnetic field its conductivity characteristics also shows a staircase function because of the Hall effect and the two-dimensional quantum layer.

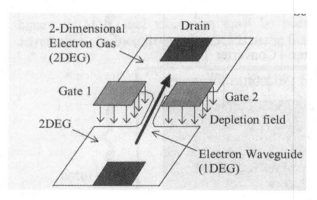

**Fig. 10.9.** Voltage impact on a channel with a two-dimensional electron gas

### 10.2.3 Electron-Wave Transistor

A further progress of the previously described devices leads to structures in which the electron waves have an even more important impact. Figure 10.10

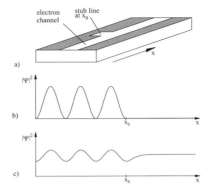

**Fig. 10.10.** An electron-wave transistor in a waveguide structure (a): (b) the channel is closed, (c) the channel is conducting

depicts a transistor in which single electrons form a current flow along the x axis.

Similar to high-frequency configurations a perpendicular stub line is connected to the channel. The principle of the device is quite simple (Fig. 10.11): The effective length of the stub line can be varied via the gate voltage. An effective length of the stub line that is equivalent to a multiple of the half-wavelength results into a short cut at the tapping. However, an effective length of the stub line that is a multiple of the quarter wave length has no impact and the electrons can pass freely through the channel.

The effective length of the stub line is altered by an external voltage. The gate electrodes of two serial-connected electron-wave transistors can serve as inputs for an AND gate.

Figure 10.12 shows the micrograph of a q-switch. The electron waves travel from the source contact, to the left or right drain contact depending on the gate voltage.

### 10.2.4 Electron-Spin Transistor

Quantum electronics can also exploit the spin of a single electron. In the literature such a magneto-electronic approach is referred to as spintronics.

In a kind of bipolar structure the emitter and collector consist of ferromagnetic layers, whereas the base is made of a semiconductor material (Fig. 10.13). As indicated by the arrows the magnetic layers have been magnetized by external magnetic fields. The spins of the electrons that enter through the emitter align to the magnetization of the emitter layer and travel towards the base. Electrons can only pass from the base to the collector if their spin aligns with magnetization of the collector. For a correct operation, the layer thicknesses have to be in the nanometric domain. All lateral dimensions can also be scaled down, which leads to a real nanoelectronic device.

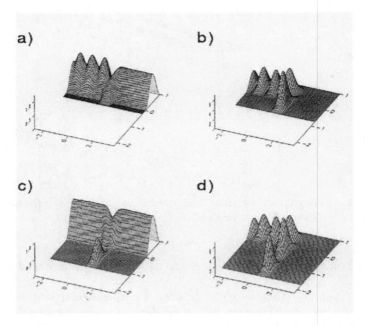

**Fig. 10.11.** Different operating conditions of the electron-wave transistor: (a) and (c) open or conducting, (b) and (d) closed or switched off

This technology of spintronics may be very promising for memories with large capacities. They may replace the magnetic hard discs.

### 10.2.5 Quantum Cellular Automata (QCA)

The idea of QCAs is quite young. Its concept describes a smart way to realize cellular automatas in a nanotechnology. Presently, cellular arrangements of quantum dots count among highly innovative concepts of solid-state electronics. They are very promising, since they depend on neither metallic wiring nor electric currents [53].

The basic QCA cell consists of four quantum dots that are arranged in a square (Fig. 10.14). Each QCA cell comprises two electrons. They autoarrange in a diagonal because of their Coulomb interaction. This results in two different configurations that are suitable for a binary representation. Via an external influence it is possible to change from one configuration to the other. Typically, the external influence consists of an electric field, meanwhile the potential barriers have to be lowered. On the basis of this principle information processing will be feasible.

Before changing the polarization of a QCA cell, the potential barriers have to be lowered in order to reduce the localization of the electrons and to in-

## 10.2 Examples of Quantum Electronic Devices

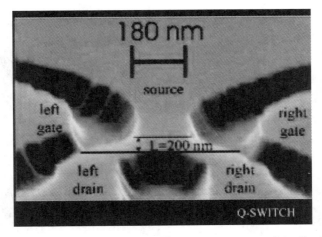

**Fig. 10.12.** Micrograph of a q-switch: Electron waves travel to the left or right drain contact

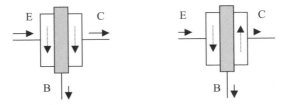

**Fig. 10.13.** Principle of the electron-spin transistor: A high collector current can only pass if the spins align with the magnetization of the collector

crease the tunneling probability. Typically an adiabatic switching scheme is applied that is similar to the adiabatic charging and discharging of capacitances (Fig. 10.15). This relatively slow switching scheme is equivalent to the clocking scheme of logic gates and supplies energy to the QCA. During the switching event the localization probability of the two electrons is uniformly distributed on the four quantum dots. After raising again the potential barriers the desired configuration is set up with respect to the external field.

One has to keep in mind that the two stable configurations depend on two electrons. These electrons have to be fed in a lateral manner into a chain or even grid QCA structure, which is not a trivial problem for any kind of technology.

The combination of single QCA elements enables the realization of relatively complex wire and gate structures (Fig. 10.16). A simple topology exists out of a linear combination that might serve as an electrical wire. In this case, the electrons interact with their neighbor cells. In the initial state all cells have the same configuration. A configuration change via an external electric field of

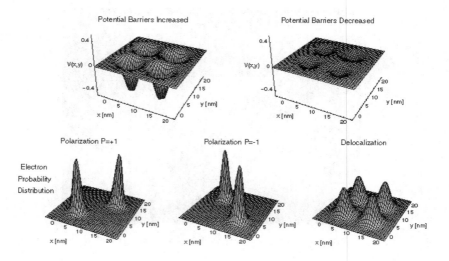

Fig. 10.14. Principle of a single quantum dot

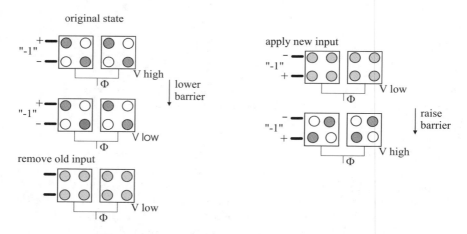

Fig. 10.15. Adiabatic switching events of QCAs

the left QCA cell evokes a configuration change of all other cells. This change assumes that the potential barriers have been lowered before. While raising again the potential barriers, the cells flip into the new configuration starting from the left side towards the right side. So information can pass without any metallic wire.

10.2 Examples of Quantum Electronic Devices   163

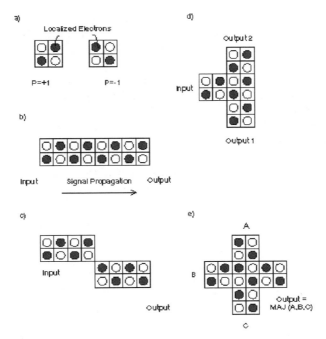

**Fig. 10.16.** Examples of basic QCA elements: (a) cell, (b) wire, (c) inverter, (d) fanout 2, (e) majority gate

According to this principle all basic elements of a computer can be realized. For example, a displacement of the cell structure results in an inverter. A T-shape arrangement offers a fanout of two.

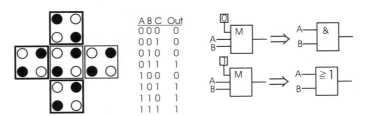

**Fig. 10.17.** Majority gate: Its realization with quantum dots, truth table, and its application as logic AND and OR gate

A cross-shaped arrangement is equivalent to a majority gate. Its output is ruled by the majority value of the inputs (Fig. 10.17). A typical application of majority gates are fault-tolerant systems. It can also serve as a universal gate, since its logic operation can be programmed. If one input is permanently connected to a fixed logic value the gate computes the logic OR operation for a permanent 1-level. For a permanent 0-level the gate performs the logic AND operation.

The example in Fig. 10.18 shows a more complex QCA gate, namely a RS flip-flop. A relatively complex clocking scheme is necessary in order to ensure the right sequence of the single operations.

At present the behavior of such quantum structures is under investigation via computer simulations (Macucci, Pisa). Up to now, it turned out that the cells are quite sensitive and that the circuits are not very robust. Furthermore, errors can only be prevented if switching takes place quite slowly in order to guarantee the termination of the transition process.

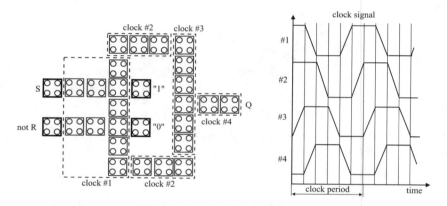

**Fig. 10.18.** QCA RS flip-flop

Additionally, other attempts try to implement such structures as solid circuits. The first promising samples in silicon have been realized on top of an $SiO_2$ layer (Fig. 10.20). Another approach couples potential wells with tunneling elements. An ambitious method tries to realize the quantum-dot arrangement within a molecule structure (Fig. 10.19).

QCAs are operated at low temperatures in order to suppress thermal influences. Room-temperature operation will become feasible if the single quantum dot is scaled below 5 nm. Such small dimensions will make it possible to store and link data in cells smaller than $25\,nm \times 25\,nm$. This packing density is close to the technology limits of solid-state nanoelectronics. However, first the influences of background charges have to be controlled.

10.2 Examples of Quantum Electronic Devices    165

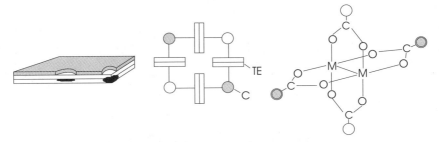

**Fig. 10.19.** Implementation of quantum-dot structures

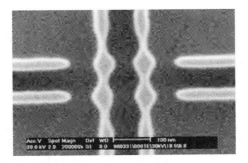

**Fig. 10.20.** Micrograph of a silicon QCA cell (University of Tübingen, Germany)

### 10.2.6 Quantum-Dot Array

A more advanced concept implements arrays of quantum dots that interact with their neighbors. Due to the coupling the quantum dots show an interesting dynamic behavior (Fig. 10.21).

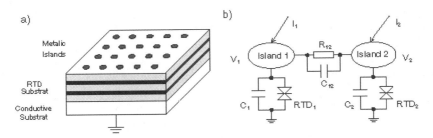

**Fig. 10.21.** Quantum-dot array: (a) technology arrangement of a quantum-dot array, (b) equivalent networking diagram of two coupled quantum dots with respect to their substrate interaction

Such quantum-dot arrays can be realized, e.g. with metallic islands such as gold clusters. They might arrange themselves in a uniform process via self-organization (Fig. 10.21). The dimensions of a single island are in the domain of a few nanometers.

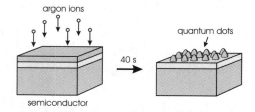

**Fig. 10.22.** Self-assembling of a quantum-dot array

This network has multiple stable states like an artificial neuronal network (e.g. a Hopfield network) which sets it apart from QCA networks. The islands can be treated as capacitances and their electrical charge can be used for information storage. A resistive and capacitive coupling between the islands is responsible for the interaction between the quantum dots. Relatively large islands result in a nonlinear, continuous network system that depends on the stored island charge. However, small islands lead to discrete energy levels due to the Coulomb blockade. The nonlinear behavior can be modeled by resonant tunneling diodes that are located between the substrate and the islands. Typical for such quantum-dot networks are image processing at low level and associative information-storage systems. Boolean operations have been already realized with small quantum-dot networks. In principle, all problems that can be described with nonlinear differential equations can be solved with these networks, assuming that the correct initial and boundary conditions have been fed into the network.

A further application might be the quantum computer: each dot represents a Qubit. If the dots are located close to each other ($\leq 4\,nm$) the electron states couple among themselves, which leads to coupled Qubits.

## 10.3 Summary

The field of quantum electronics has given rise to many interesting approaches. Whether they are useful or not for nanoelectronics has to be checked in the future. Most of these approaches are based on mesoscopic structures and exploit the wave properties of the electrons.

Quantum-dot structures show interesting concepts with typical nanometric dimensions of less than $5\,nm \times 5\,nm$. However, for their successful operation as

QCAs complex wiring schemes are indispensable. Such arrays are also useful for quantum computing.

# 11
# Bioelectronics and Molecular Electronics

Nanoelectronics has various ways to go further. Presently, nanoelectronics moves along three main directions (Fig. 11.1):

1. Solid-state nanoelectronics: Within very small crystal structures electrons are primarily ruled by their wavefunctions. Solid-state nanoelectronics follows a top-down approach and it comes from the area of system implementation. Such concepts have been discused in the previous chapters.
2. The specific states of the electrons in a molecular structure have to be considered when it comes to molecular electronics. Molecular electronics is a bottom-up approach and it starts from questions like: What can be realized with such basic elements?
3. A further approach is motivated by biology. Therefore, it is called bioelectronics or it is referred to it as wet electronics. Bioelectronics is inspired by biology and it tries to copy certain concepts of biology. Such an approach has been described in Chap. 4.

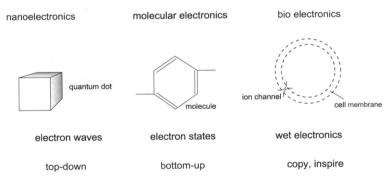

**Fig. 11.1.** Rough classification of nanoelectronics

This chapter can only be a sort of summary of approaches to bioelectronics and molecular electronics. At present, developments in these fields take place quite rapidly so that any more complex explanation would be out of date very soon. Therefore, this chapter only claims to give a coarse overview on basic principles.

## 11.1 Bioelectronics

The membrane is a very important part of a biological cell. It can adjust its conductivity via ion channels over several orders of magnitude. A possible approach of molecular electronics might be the reproduction of such biological cells and use them as information-processing units. A substantial step would be the realization of an artificial membrane.

Langmuir-Blodgett developed a simple method to realize artificial membranes (Fig. 11.2): A hydrophobic substance that has been added to water aligns as a monolayer on top of the water surface. The aligned molecules attach to a substrate via adhesion if dipped in water with the monolayer. As the substrate is pulled out a second molecular layer attaches to the first molecular layer. The second layer has the inverse orientation, which results in a molecular double layer. This double layer is quite similar to a cell membrane. For certain, it is not easy to process this thin cell membrane.

Ion channels can be integrated into the double layer, just like biological cells (Fig. 4.6). Destinct molecules can activate these channels and turn them into a conducting state. This behavior is equivalent to a controllable switch.

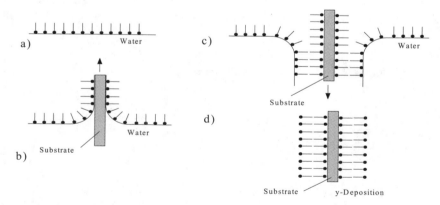

**Fig. 11.2.** Realization of artificial Langmuir-Blodgett films: (a) molecules align with their hydrophobic side on top of a water surface, (b) the aligned molecules attach to a substrate via adhesion, (c) and (d) by means of a further process step a double layer can be realized

The next step is to establish a network out of these switches that is suitable for information processing.

### 11.1.1 Molecular Processor

A further application of the artificial membrane is the biomolecular processor. The objective of molecular electronics is to process nontactile information such as light and electrical signals via enzymes. Figure 11.3 shows, for instance, a tactile molecular processor. Receptors in the input layer transform nontactile information into tactile signals. If the tactile signals match with the enzymes that are located in the output layer the specific output is activated. The program (software) of the molecular processor depends on the receptor molecules and the read-out enzymes. Just like DNA structures, the molecules have to match to each other. This concept can be extended to a teachable and self-reproducible processor. It might consist of various series-connected chambers that are responsible for distinct tasks. Certain principles of a tactile processor can also be found within the synapses of biological networks that have been described before.

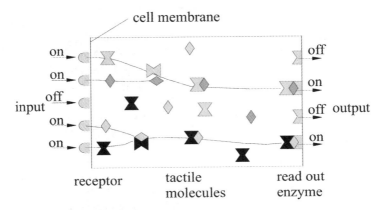

**Fig. 11.3.** Basic principle of a tacticle molecular processor

The molecules act as data carriers. They transport information without consuming practicaly any energy. Their movement is due to the Brownian molecular movement, which anyhow exists because of the thermal energy. This gratis information transport is an important issue for very large systems. In this connection the following relation of Einstein is of fundamental relevance:

$$< r^2 > = \alpha \cdot t = 6kT\frac{t}{\mu_t}. \tag{11.1}$$

According to it the mean square distance r between a particle and its origin increases with the time t.

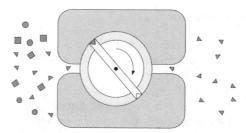

**Fig. 11.4.** A micromotor drives a rotor that sorts molecules, such a system might be applied within the tactile processor, according to Drexler

The factor $\alpha$ depends on the Boltzmann constant k, the temperature T, and the coefficient of friction $\mu_t$ that is proportional to the diameter of the particle. The relation describes the area of influence for a single particle. The more time that passes the higher the probability that the particle has arrived at its destination. Obviously, molecular processors that are based on this principle operate quite slowly. In this connection, highly parallel structures compensate this drawback, as has been already explained for the DNA computer. Another approach tries to overcome this disadvantage by designing mechanical molecular structures. Drexler and others design nanomechanical computers that are based on the same principles as the first mechanical computers 100 years ago. A further approach is a rotor that only transports specific molecules (Figs. 11.4 and 11.5). According to the shape of the molecules the rotor sorts out specific molecules.

The studies of Drexler and Merkle [54] describe interesting nanomechanical approaches. However, from the present point of view it is unlikely that future nanocomputer technology will be based on such structures. Nanomachines like the DNA computer seem to be more promising candidates even if they are far in the future.

Probably, a biomolecular computer needs more care from the user than an ordinary PC. From this point of view the Tamagoccis that came into fashion in 1998 can be considered as a huge field test. It proves that users can really be expected to care about their computers.

### 11.1.2 DNA Analyzer as Biochip

An interesting application for the well-being of mankind are analytical chips that might identify certain diseases and serve for food inspections since the DNA carries the whole genetic information of all living creatures. Within such minilabs nanostructures are of fundamental relevance for fluidics. Additionaly nanomechanical structures like nanovalves and nanopumps have to be integrated (Fig. 11.6). These means make it possible to transport substances to their destinations on the chip and replace a microspotter.

For the analysis itself different methods are feasible. The proteins in question get selected by DNA truncks. Their properties can be discovered by the fluorescence behavior of their molecules. In the long term, pure electronic methods are preferable to minimize the dimensions of the minilab and to avoid complicated optical setups. This case requires a number of electronic sensors to be operated in parallel (Fig. 11.7).

A single sensor electrode within the array consists of an interdigitated gold layer. Using a microspotter or a nanofluidic system, single-stranded DNA probe molecules are spotted and immobilized at the gold surface. After immobilization, an analyte containing target molecules to be detected is applied to the whole chip and hybridization occurs in the case of matching DNA strands. For read-out, a redox-cycling-based electrochemical sensor principle is used: After a washing step, a suitable chemical substance is applied and electrochemically redox-active compounds are created by an enzyme label bound to the target strands. Applying simultaneously an oxidation and a reduction potential to the interdigitated gold electrodes, a redox current between these electrodes occurs whose magnitude depends on the amount of double-stranded DNA at this sensor position. The biochip provides the results of the tests performed directly in the form of electronic signals.

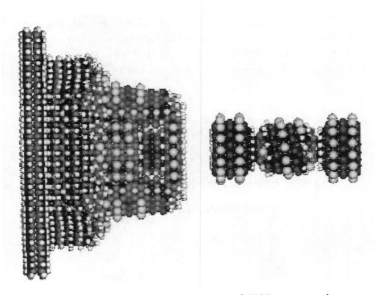

**Fig. 11.5.** Example of a nanomachine, it consists of 6165 atoms and acts as a pump for atoms; courtesy of Institute for Molecular Manufacturing, www.imm.org

## 11.2 Molecular Electronics

### 11.2.1 Overview

Molecules are local objects that can take different states. Observable features of these states can be used for information storage. Figure 11.8 depicts different possibilities for a binary information storage. A molecule with a number of unpaired spins can be aligned in a parallel or antiparallel state to the preferred spin condition. These spin conditions can be detected directly by their magnetic interaction or by optical means. Other binary sytems consist of molecules with hydrogen bonds that can be described, in general, as double minimum potential DMP (Fig. 11.8b). The proton can take two stable postions A and B. The two resulting dipole moments might represent the two logic values 0 and 1.

Molecules often have two different configurations, like 'trans' and 'cis', as is depicted in Fig. 11.8c. A reversible configuration change might also serve as a binary memory. Different absorption properties can be used for distinction purposes. Even more interesting seems the complex of a donor-acceptor-bridge (Fig. 11.8d) in which an electron is transfered from the donor to the acceptor. The donor is in the reduced state before the transfer and afterwards in the oxidized state. Both states can be used for a binary representation.

Figure 11.9 illustrates a fundamental element of molecular electronics. An electrical signal controls a bistable molecule via a molecular wire. The semiconductor has very small dimensions and consists of a few molecules so that the band theory of solids is no longer applicable. At present, it is very difficult to switch molecular devices with electrical signals, also the read-out process has not been realized by electrical means. At this early stage of molecular electronics switching is still done by light and the read-out process is established,

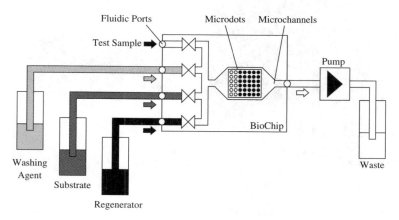

**Fig. 11.6.** Setup of a microanalytical chip

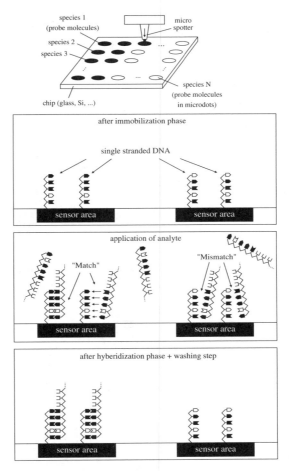

**Fig. 11.7.** General operational principle of DNA chips

for example, by spectroscopy. Obviously, complex information processing units are not yet feasible.

Examples of bistable molecules are salicyl-aniline and viologene, whereas conducting polymers might serve as molecular wires. In this case information transfer depends on a soliton. An acryl plastic layer is put on top of a protein layer. Then the acryl plastic is sliced into nanometric wires by an electron beam. This technique will be developed further in the future.

### 11.2.2 Switches based on Fullerenes and Nanotubes

Carbon naturally appears as graphite and diamond. Additionally, it can also be found in clusters of 60 carbon atoms. In general, these clusters are termed

# 11 Bioelectronics and Molecular Electronics

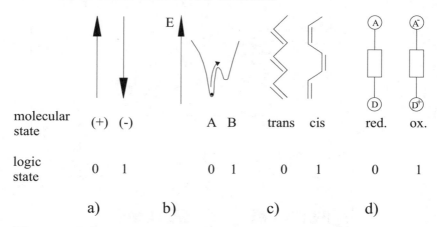

Fig. 11.8. Different molecular memory structures, according to C. Mehring

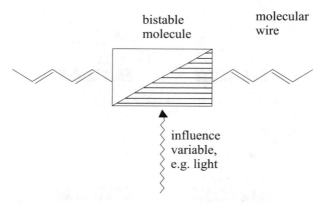

Fig. 11.9. Basic molecular electronics element: Often light is used as input signal, which is not suitable for complexer units

fullerenes or C60 molecules (Fig. 11.12). Their conductance depends on their deformation. The current that can flow through a macromolecule increases for higher pressures (Fig. 11.10). Similar to the tunneling effect, the band structure is shifted, which effects the conductance. Piezoelectric actuators can be applied for deformation purposes.

The first designs integrated macromolecules into conducting paths. A piezoelectric actuator applies a pressure on a single molecule (Fig. 11.11). This semi-mechanical transistor can be used as a switch. The active region of the composition is limited to the macromolecule, but the entire approach suffers from the relatively huge and complex actuator. Therefore, for the time being this switch is nothing more than a vision that lacks a VLSI concept.

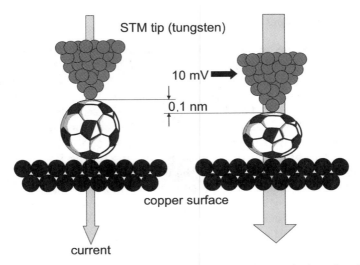

**Fig. 11.10.** Molecules as controllable resistors or transistors: A piezoelectric actuator alters the resistance of a C60 molecule that is located between to metal contacts

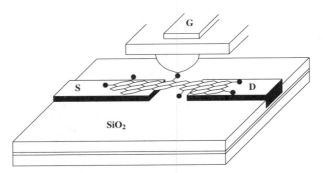

**Fig. 11.11.** The carbon-tube transistor with source, gate, and drain

It is possible to design and simulate basic circuits that are based on these transistors. A spike capacitance of $10^{-15}\,F$ and a molecular resistance of $50\,M\Omega$ result in a relatively slow inverter behavior with a time constant of approximately $100\,ns$. However, memory densities can reach magnitudes far beyond Si-DRAM.

Nano tubes (Fig. 11.12) offer a similar vision for switches. At present, it seems as if this technique might be used for a straightforward implementation of nanotransistors.

### 11.2.3 Polymer Electronic

Polymers become more and more important for electronics in terms of electroluminescent displays and batteries. They are also integrated into transistors (Fig. 11.13). Conducting and nonconducting regions can be realized via an exposure of a PANI layer. The conducting regions can be used to define source, drain, and gate contacts. The channel is made of the semi-insulating polymer PTV where as PPV suits as the gate insulator. The fabrication of such transistors is very inexpensive since they can be integrated on flexible plastic sheets. Circuits that are based on this technology are interesting for both wearable electronics and for a replacement of the well-known bar code.

Nanoelectronics might offer possibilities to fabricate switches and connections that are based on polymers (J.C. Ellenbogen). Depending on the individual polymer structure conductors and insulators can be established (Fig. 11.14). The idea of polymers for wiring purposes already appeared in Chap. 9.

Table 11.1 compares the main features of polymers, semiconductors, and metallic conductors. At first glance, the poor values of the polymer materials are quite impressive. If we scale down the values into the nanometric regime and assume single-electron transportation the polymer approach catches up with the conventional technologies.

Logic circuits require nonlinear device characteristics. The basic structures in Fig. 11.14 lead to the molecular diode in Fig. 11.15. The polymer struc-

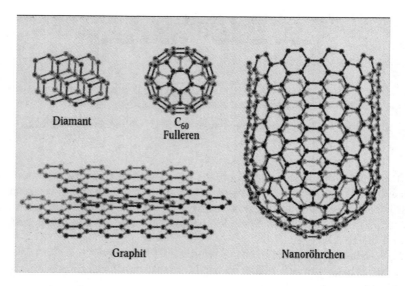

**Fig. 11.12.** Besides graphite and diamond carbon also appears as fullerenes and nanotubes

## 11.2 Molecular Electronics

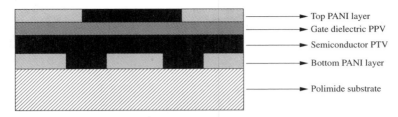

**Fig. 11.13.** Polymer transistor for most inexpensive electronics, e.g. wearable electronics

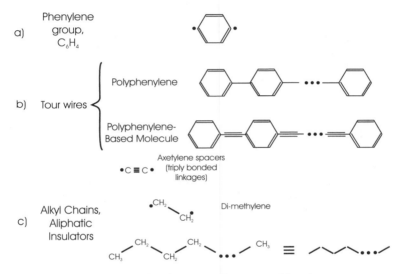

**Fig. 11.14.** Molecules as conductors and insulators

**Table 11.1.** Comparison between polymers, semiconductors, and metals

|  | Polymer | Nano carbon tube | Semiconductor | Copper |
|---|---|---|---|---|
| Typical voltage [V] | 1 | 1 | 0.1 | $10^{-3}$ |
| Typical current [A] | $10^{-8}$ | $10^{-7}$ | 0.01 | 1 |
| Current density [$electrons/nm^2$] | $10^{12}$ | $10^{11}$ | $10^6$ | $10^6$ |

ture is located between two gold contacts. Gold seems to be very suitable for interconnects between different blocks.

A logic function like an AND gate can be realized with two diodes and a resistor (Fig. 11.16). This circuitry is strongly related to diode-transistor

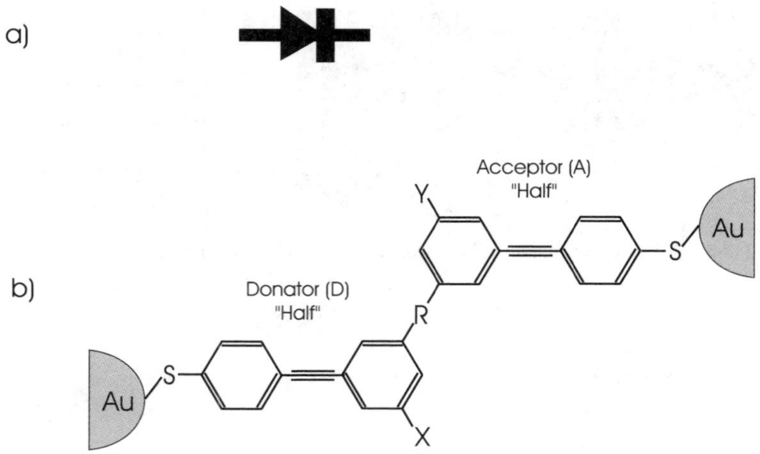

**Fig. 11.15.** Molecular structure of a diode

logic. In principle it is possible to make up larger chips, however, it is unlikely that this concept will succeed.

Using this concept, even more complex units are feasible [55]. The discrete molecular switches within the logic gate are triggered via polymer wires. The direct transformation of conventional logic families into the molecular domain appears not to be reasonable, e.g. power that is dissipated during switching is more difficult to eliminate. A possible solution is the concept of the *Fredkin logic* that avoids the loss of information within the gate.

### 11.2.4 Self-Assembling Circuits

The three-dimensional molecular structures might be put together like a puzzle. In general, the active molecules suffer from mechanical weakness so that a straightforward realization of a three-dimensional system is not feasible. Therefore, a substrate has to be fitted to the molecular structure for stability and insulating reasons.

In the following, the basic idea of a technology will be explained that is still in its inital state. As an interesting example we will take a closer look at self-organizing cells that are located in a zeolite crystal. The objective of the crystal is to define the mechanical structure. Some zeolite crystals have a rectangular channel structure, in the nanometric regime (Fig. 11.17). It is possible to locate semiconductor dots in the cross-points of the channel structure whereas polymer wires can be placed into the channels. The semiconductor dots serve as memory cells, since they show a bistable behavior.

This example shows that molecular electronics offers the possibility of a three-dimensional integration. Even for two-dimensional layouts the packing

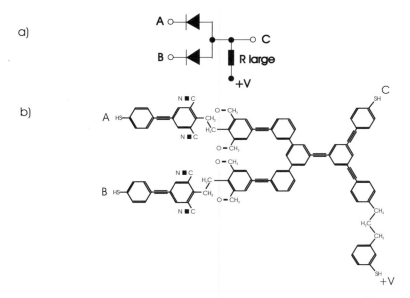

**Fig. 11.16.** Polymer structure of a logic AND gate: By reversing the diodes' polarity and by applying a negative supply voltage the gate turns into a logic OR gate

density can be increased with this technology by a factor of four. By using all three dimensions the capacity will increase tremendously. For instance, a rectangular channel structure with a grid of $10\,nm = 100\,A$ and a volume of $1\,cm^3$ has $10^{18}$ cross-points. In principle, each cross-point can be used for information storage, assuming the controllability.

Obviously, for these fine structures the question arises as to how they can be organized. Lithography means are not appliable for three-dimensional systems. Self-organization approaches offer a viable structuring concept. One has to differentiate between self-organization and self-assembling . Self-assembling refers to crystal growth or epitaxy, whereas self-organization can be found, for example, in neural networks, which have been explained in Chap. 7. By electrical means, data has a direct influence on the physical structure that leads to a self-organization. According to an external blueprint the required circuitry is generated, very similar to biological systems.

Later, molecular electronics fills the gap between nano- and microstructures as an interface. For experimental purposes junctions can be made via the scanning tunneling microscope (STM), as described in the following chapter. For more complex connections, as appear in dense systems, lithography has to be developed further.

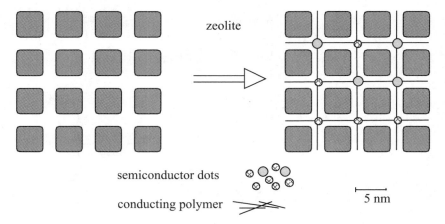

**Fig. 11.17.** Three-dimensional zeolite memory: Semiconductor dots are located in each cross-point, whereas polymers in the channels serve as wires

### 11.2.5 Optical Molecular Memories

Figure 11.18 reveals a molecular shift register with optical clock distribution. Within each period the stored information moves to its neighbor. It is problematic to deal with this optical clock distribution when it comes to more complex systems.

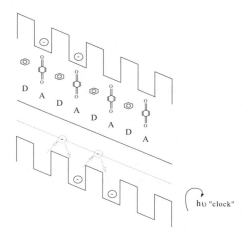

**Fig. 11.18.** Molecular shift register with optical clocking, according to Haarer

Biomolecules offer a fast and dense memory storage. Primarily they are very well suited for three-dimensional architectures and parallel data processing. For the time being such memories can only be realized as optical systems since the reactions within the biomolecules depend on light. Bacteriorhodopsin is an interesting molecule for data storage. This protein can absorb light, meanwhile it changes its internal states. According to the color of the light bacteriorhodopsin takes distinct states. These different states are easily detectable so that bacteriorhodopsin is an appropriate switch and memory element.

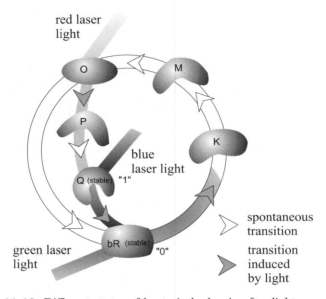

**Fig. 11.19.** Different states of bacteriorhodopsin after light exposure

During light exposure bacteriorhodopsin passes through a photochemical cycle, which results in structural changes of the molecule (Fig. 11.19). Green light transforms the protein from its inital state bR into K. From the K state the protein changes spontaneously into M and then into the O form. Without any further influence the protein changes back to its inital state bR. A preexposure of red light leads the protein to the P state, from where it changes rapidly into the very stable Q state. The Q state can only be left via bluelight exposure to the inital state. The two states bR and Q are suitable for the binary memory representations 0 and 1. In the following example the optical memory depends on these two states of the bacteriorhodopsin molecule.

The intermediate states P and Q result from the sequential absorption of a green and a red photon. They are suitable for memory storage. The memory storage itself takes place both in parallel and in three dimensions. The media has a cubic shape and consists out of bacteriorhodopsin. The cube can be

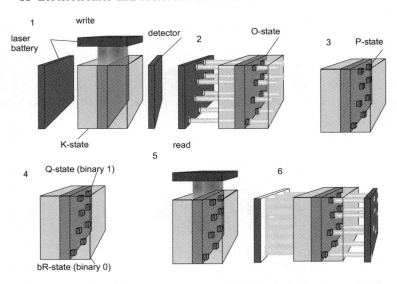

**Fig. 11.20.** Optical memory based on the bacteriorhodopsin molecule

exposed with batteries of lasers from two different directions (Fig. 11.20). The upper laser battery emits green light and starts the photochemical process of the bacteriorhodopsin molecules for a selected layer. After a few milliseconds the concentration of molecules that are in the O state has increased. Red laser light that comes from the left side can now write at distinct places a logic '1'. These molecules take the P and then the stable Q state, whereas the rest of the molecules return to the initial state. Blue light can reset the whole memory block. The read operation depends on the selective absorption of red light via the intermediate O state. First, the selected layer has to be exposed with green light, so that the molecules in the initial state start with the photochemical cycle. After two milliseconds the read-laser battery has to emit faint light. Molecules in the Q state are transparent for the red light whereas molecules in the bR state appear to be opaque for the red light because the previous pulse of green light had put them into the O state.

A detector registers the translucent light and reads the logic states. For one layer the read operation takes place in parallel. Therefore, the three-dimensional structure results in a short access time, a high memory bandwith, and a high storage density. Hopefully, one trillion bits per $cm^3$ are feasible.

At present the objective is not to realize a whole computer within this technology. First, memory applications that exist of biomolecules will be realized and will be connected via a technology-invariant interface to the computer system. Initially, single data will be stored in several molecules and in a further step towards even higher storage densities a single bit might be stored in one molecule.

## 11.3 Summary

The suggested approach that tries to copy biological systems leads to interesting concepts. Up to now these concepts have almost no impact for today's applications. This might change if a complete analytical minilab can be integrated into a single chip or if huge memories can be provided.

In terms of molecular electronics carbon structures and polymers are the most promising candidates. Applications that are based on huge mass-storage systems probably will be rapidly revolutionized by optical molecular memories.

# 12
# Nanoelectronics with Tunneling Devices

From today's point of view tunneling elements (TE) such as the resonant tunneling diode (RTD) are of fundamental relevance for nanoelectronics. They are usually of nanometric scale in a single dimension and are promising candidates as precursors for future nanoscaled ULSI circuits. At present they are the most mature type of all quantum-effect devices . Compared to single-electron transistors (SET) and more advanced quantum-dot architectures, resonant tunneling devices are already operating at room temperature. Technological advances, such as the development of a III-V large scale integration process, and the demonstration of a $Si/Si_{0.5}Ge_{0.5}/Si$ resonant interband tunneling diode, are a challenge for circuit designers to develop digital logic families and memory arrays.

## 12.1 Tunneling Element (TE)

Among all mesoscopic switching elements the tunneling elements (TE) are at present of fundamental relevance for nanoelectronics. A tunneling element consists of two conducting materials separated by a very thin insulator. By means of bandgap engineering we can tune the I-V characteristics of the tunnelling element in such a way that it has a region with negative differential resistance (NDR). The tunneling diode (TD) and the resonant tunneling diode (RTD) are the most common tunneling elements.

The basic idea behind resonant tunneling device circuit design is to exploit the nonlinear I-V characteristics with the typical negative differential resistance (NDR) region. In this context the isolation of the gate input and output signals is of fundamental relevance and has motivated the implementation of different three-terminal tunneling devices. The common feature of these different devices is the combination of electronic amplification and NDR. In addition to the technologically oriented research a further prerequisite for nanoscale integration is the investigation of suitable logic families, architectures and the development of a design framework.

### 12.1.1 Tunnel Effect and Tunneling Elements

Figure 12.1 shows the schematics of two basic tunneling element configurations. A convenient solution is the TE with a vertical layer structure (Fig. 12.1a), but technology most commonly uses horizontal layer structures. These layers can be precisely deposited and are in the range of 1 to 5 nm, which roughly is equivalent to the number of $SiO_2$ layers.

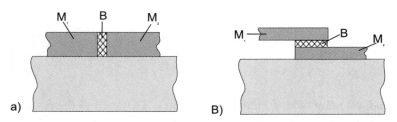

**Fig. 12.1.** Schematic views of tunneling elements with vertical (a) and horizontal (b) oriented barriers

The tunnel effect refers to particle transport through a potential barrier where the total energy of a classical particle is less than the potential energy. In the classical sense this particle transport is not possible. This effect can be explained if the particle is treated as a material wave. The course of the $\Psi$ function in the x direction can be described by Schrödinger's wavefunction. As an example Fig. 12.2 shows the solution for a specific case. The solution of the wavefunction must satisfy certain boundary conditions at the abrupt interfaces of the barrier. This leads to a certain portion of an incident wave being transmitted and a certain portion reflected. On the left side of the barrier one can observe the entering and the reflected wave. Within the barrier the wave is attenuated, whereas on the right side of the barrier the wave of the tunneled electron can be found. The wavefunction has to be continuous and differential and can be described by four amplitude values and four equations. Figure 12.3 depicts the solution of these equations and is equivalent to the tunneling probability of the electrons. The tunneling probability increases for higher particle energies.

Due to reflections of the particle wave not all electrons can cross the barrier, even if their energy is higher than the potential energy of the barrier (Fig. 12.3). The electron wave can only cross the barrier without dissipation if the energy of the electrons is six times higher than the potential energy of the barrier.

For the lower part of the function in Fig. 12.3 there is an approximation: The amplitudes of the $\Psi$ function on both sides of the barrier are proportional to the probability of presence of the particles. The ratio of these amplitudes approximates to the tunneling probability D:

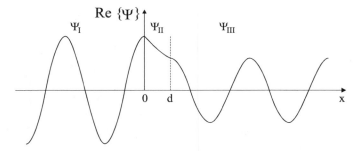

**Fig. 12.2.** Tunneling effect: The material wave is attenuated within the barrier

$$D \approx C\left[-\frac{d}{\hbar}\sqrt{2m(W_0 - E)}\right]. \qquad (12.1)$$

A particle that has energy E can tunnel with probability D through a potential barrier that has height $W_0$ and width d. The aperture probability D is higher for lower and thinner potential barriers. C is a constant proportional factor, h is Planck's constant and m represents the electron mass. The portion of tunneled particles as well as the tunnel current can be gathered from the tunneling probability. This effect is exploited among other things within the tunneling element (TE) which can be approximately modelled as an ohmic resistor.

An electric field distorts the barrier shape, as denoted in Fig. 12.4. The upper part of the barrier gets smaller so that the tunneling probability increases.

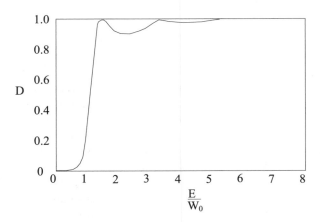

**Fig. 12.3.** Tunneling probability as a function of the normalized energy, according to the Schrödinger equation

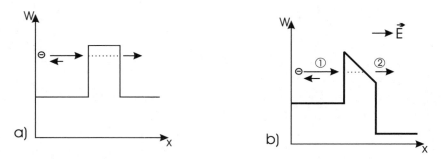

**Fig. 12.4.** Example for the tunneling process: a) tunneling barrier in its ground state, b) tunneling element with applied voltage.

Once the electric field is removed the barrier shape returns to its inital state and the tunneling probability decreases. This effect is used, for instance, in electrically erasable programmable read-only memories (EEPROM).

Tunneling elements are very interesting switching elements for nanoelectronic applications because the electron transport takes place without any loss of energy. Additionally, the switching speed of the elements is very high, since the potential barriers are very thin. Physicists discuss whether the transport mechanism in the TE is faster than the speed of light. To what extent this assumption might be correct and what impact it might have on nanoelectronics remains, for the time being, an open question. A TE can also be a source of errors: If we think about a MOS transistor with a very thin gate oxide the tunneling current through the gate oxide has to be taken into account, the same effect can also occur between conductors that are separated by a very thin insulator.

### 12.1.2 Tunneling Diode (TD)

The interesting feature of a tunneling diode is its negative differential resistance. It consists of two semiconductors that are separted from each other by a thin insulator. Figure 12.5a illustrates the potential barrier caused by an insulator. Electrons can tunnel through this barrier only if it is thin. A further requirement for the tunneling process is the existence of an unoccupied band on the other side of the barrier. The unoccupied band can pick up the tunneling electrons (Fig. 12.5b). The barrier thickness can be altered by an electric field. The tunneling process can still be impossible, even if the barrier thickness is very thin because of the absense of a free band at the other side of the barrier. For higher electric fields the influence of the barrier can be neglected and the common diode effect can be observed. The complete I-V characteristics of the tunneling diode with its negative differential resistance region is depicted in Fig. 12.6.

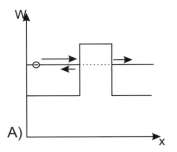 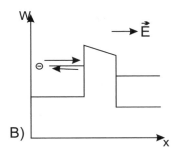

**Fig. 12.5.** Tunneling effect in semiconductors: (a) Within material that shows a band structure tunneling is only possible if a free band is available on the other side of the barrier, (b) an electric field can displace the band structures so that tunneling becomes impossible

The shape of the I-V characteristics reveals the underlying physical effect: One can observe a falling slope in the characteristics, which in general is referred to as the *negative differential resistance* (NDR). The essential parameters of the TD are the peak current $I_P$, the valley current $I_V$, the peak voltage $V_P$, and the valley voltage $V_V$. These parameters can be derived from the band diagram. The ratio of $I_P$ to $I_V$ is important for circuit designers because of its direct influence on the signal amplitude. In this context the minimization of the valley current is a challenging issue. In the quiescent state of the gate this current has still to be delivered. Therefore it seems that this technique is only suitable for low-power applications if the valley current can be reduced to approximately zero. For very fast circuits the valley currents can almost be ignored, since the (dis-)charging of parasitic capacitances is the main cause for power dissipation.

For many years the realization of the thin insulator has been a crucial issue and has been solved by a technological trick: Very high doping levels result in a degradation of the semiconductor in the sense that its band structure overlaps. The Esaki diode consists of a pn-junction with extremely high doped semiconductor regions (Fig. 12.7b). The upper part of the valence band that is located in the p-type region overlaps with the lower part of the conduction band that is situated in the n-type region. Both regions are separated by a very thin space-charge region that should prevent (in the classical sense) any current for small voltages. Quantum mechanics explains the tunneling of electrons through such a thin space-charge region. Therefore one can observe a high current through this region even for small voltages (Fig. 12.7b). This current reaches a maximum value for a specific voltage. For higher voltages the current starts to drop because the overlap of the above-mentioned bands shrinks. For even higher voltages the regular diode current starts to contribute to the overall current, which therefore increases for a second time. Thanks to this technological trick it is possible to obtain an I-V characteristics with a

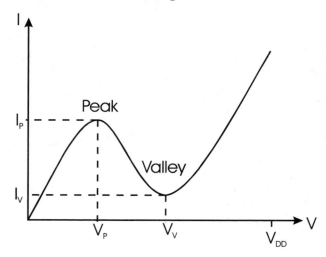

**Fig. 12.6.** Typical I-V charateristics of a tunneling diode

negative differential resistance region. In a first-order approximation the I-V characteristic can be described by the equation of van der Ziel:

$$I = A'' \frac{V(V_V - V)}{V_T} + I_{S0}\left(e^{\frac{V}{V_T}} - 1\right), \quad V_V > V > 0. \qquad (12.2)$$

$A''$ is a constant, $V_V$ symbolizes the valley voltage, and $I_{S0}$ stands for reverse saturation current.

Among all tunneling elements the tunneling diode is the most popular one and was realized for the first time in 1968. It has been implemented in silicon and did not have any important impact on microelectronics. Only when new integration forms like the resonant tunneling diode (RTD) were developed did the tunneling diode start to become successful. The resonant tunneling diode has several degrees of freedom, so that the I-V characteristics can be adjusted in a wide range.

### 12.1.3 Resonant Tunneling Diode (RTD)

The typical negative differential resistance of the tunneling diode is very noticeable for the resonant tunneling diode (Fig. 12.8). Within the RTD the source and drain contacts are seperated from the channel region W by tunneling elements. The channel region W can be described as the potential well.

Its behavior can be derived from the Esaki tunneling diode. The band structure of the channel has to be approximately on the same level as the band structure of the source contact. Otherwise only a small current flows through the device.

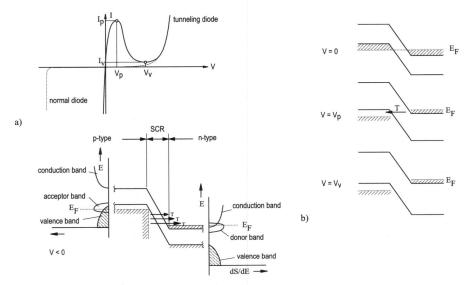

**Fig. 12.7.** Tunneling diode as a solid-state switch: I-V characteristics with NDR (a) and its band diagram (b)

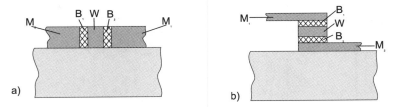

**Fig. 12.8.** Cross section of the RTD: (a) with vertical and (b) with horizontal barriers

A serial combination with an ohmic resistance shows three operating points (Fig. 12.9): $A_1$ and $A_3$ are stable, whereas $A_2$ is a metastable operating point. The stable operating points can be used to store data, whereas the metastable one is intersecting for very sensitive applications such as amplification. For example, a faint external signal may delegate the gate to one or the other stable operating points. The switching takes place in a very fast way because of the thin barriers.

The term 'resonant' refers to the behavior of the de Broglie wave in the quantum well. It is located between the two barriers and emerges if the energy of the tunneling electrons is equivalent to the level in the quantum well (Fig. 12.10). Under these conditions the transmission T also reaches its highest peak. The width of the transmission chraracteristics results from the interac-

194    12 Nanoelectronics with Tunneling Devices

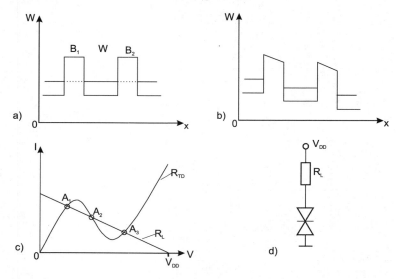

**Fig. 12.9.** Resonant tunneling diode: (a) band diagram for $V_{DD} = 0$, (b) band diagram for $V_{DD} > 0$, (c) I-V charactistics with three operating points, (d) RTD with load resistance $R_L$

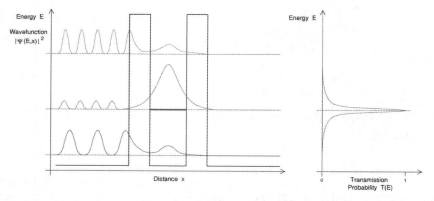

**Fig. 12.10.** Resonant effect: The material wave reaches its maximum if the energy of the tunneling electrons is equivalent to the level in the quantum well

tion of the electrons with the crystal lattice. The electrons can emit or receive energy from the crystal lattice so that the energy condition still holds if the energy deviates slightly from the level in the quantum well.

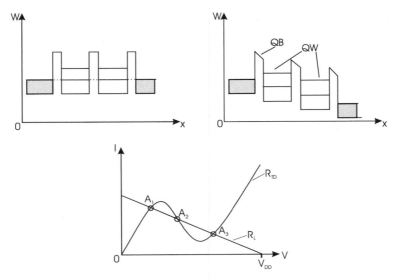

**Fig. 12.11.** RTD with two wells and three barriers

A different implementation form of the resonant tunneling diode uses three barriers (Fig. 12.11). In this case the typical I-V characteristics depends on the displacement of the two quantum wells. In contrast to the previous RTD type this implementation does not need any specific band structure at the drain and source contacts.

Such a device can be made out of thin metal layers with even thiner insulator layers ($1-2nm$) that act as quantum wells. Depending on the quantum well different configurations of energy levels are attainable (Fig. 10.4).

In general, the tunneling diode has very good switching properties, but suffers from the drawback of being a two-terminal device. The isolation of the gate input and output is of fundamental relevance and cannot be achieved by the tunneling diode on its own. This has motivated the implementation of different three-terminal tunneling devices. The common feature of these different devices is the combination of electronic amplification and of the effect of NDR. The monolithic integration of the tunneling diode with three-terminal devices is a prerequisite for manufacuturing such circuits. These devices, namely field effect transistor and the bipolar transistor offer the required input-output isolation and gain.

## 12.1.4 Three-Terminal Resonant Tunneling Devices

The switching behavior of two-terminal devices like the tunneling diode can not be controlled via a third terminal. Without this essential feature no isolation between the gate's input and output is feasible. Nevertheless, this isolation is of fundamental relevance when it comes to circuit design.

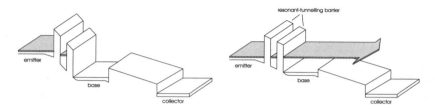

**Fig. 12.12.** Band diagram of the RTBT

The monolithic integration of an RTD structure into the emitter branch of a bipolar transistor is called a 'resonant tunneling bipolar transistor' (RTBT) which offers a solution for this scenario. In this case, the regular pn-junction between the two terminals *base* and *emitter* is replaced by an RTD structure (Fig. 12.12). The I-V characteristics (Fig. 12.13a) reveal the amplified collector current with the typical NDR behavior.

Another solution for the isolation problem is the combination of the field effect transistor (FET) with an RTD. Serial and parallel combinations of these two devices show the negative differential resistance (Fig. 12.13b). In both cases the I-V characteristics can be altered by the gate voltage of the FET.

Figure 12.13c shows the schematic of a single-electron transistor (SET) that consists of small capacitances and two tunneling elements. Again we can observe NDR regions in its I-V characteristics that are typical of all quantum-effect devices. A closer look at the SET will be taken in the next chapter.

## 12.2 Technology of RTD

According to the schematic views of an RTD one would expect two layers of semiconductor material that form the quantum wells and one additional layer for the quantum well. Actually, an implented RTD shows this structure, but in a more complex surrounding. To achieve a stable assembly of these thin layers a relatively complex layer structure is necessary. The example in Fig. 12.16 shows a typical RTD structure within the indium gallium arsenide technology. One can anticipate the challenging implementation of such a device.

Figure 12.14 shows different implementation forms of the RTD. Its combination with a FET can be realized in series (a, b, e) as well as in parallel (d, c, f).

## 12.2 Technology of RTD

The main objective of the applied research in this field is to implement such RTDs in a silicon or silicon-germanium technology and to transform the already attained knowledge into the mainstream technology. The experiences of the silicon technology have been gathered over many years and it may be feasible to exploit these experiences onwards. Later, RTDs could be easily co-integrated with conventional CMOS technology. This step would accelerate the introduction of the RTD technology and further nanoelectronic components.

In Fig. 12.14a and e the tunneling diode consists of an Esaki diode. The diode is composed of an $n^+$-doped contact and a $p^+$-doped diffusion region.

Figure 12.15 depicts the micrograph and the cross section of an RTD-FET combination that has been realized in an indium phosphide technology. The micrograph reveals the present drawbacks of this approach: The RTD can be perfectly implemented as a mesoscopic device, whereas the remaining components such as wiring and contacts are still too large for any nanoelectronic application. Thus, negative aspects are the low packing density as well as the large parasitic capacitances that increase gate delay times. In this connection the question arises if this approach is viable under these conditions.

This problem might be solved if it is possible to introduce RTD into the polymer technology. In Fig. 12.17 an RTD is composed of insulating and conducting molecules that form the tunneling elements. But nevertheless, the question of how to build complex circuits and architectures out of these mechanically unstable elements remains open.

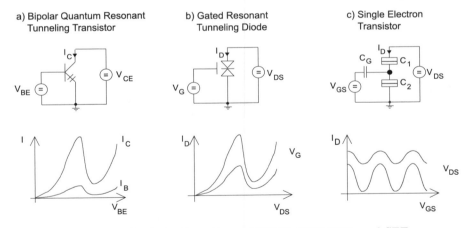

**Fig. 12.13.** Basic configurations of RTBT, FET-RTD, and SET

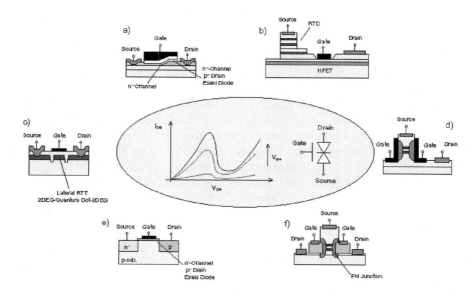

**Fig. 12.14.** Some FET-RTD combinations

## 12.3 Digital Circuit Design Based on RTDs

### 12.3.1 Memory Applications

There are already many proposals for memory applications and logic circuits that make use of resonant tunneling diodes. Figure 12.18 illustrates a static memory cell that is composed of serially connected RTDs and a FET.

The FET selects the RTD combination for read and write access. During the write operation the signal on the bit line drives the memory node to the desired state. As we will see in the following, the switching operation is faciliated if the supply voltage is clocked.

### 12.3.2 Basic Logic Circuits

In the style of the classic CMOS circuitry RTD-based logic gates can be composed: Figure 12.19a shows an inverter. The tunneling diode provokes a hysteresis during the switching operation. This is not a parasitic side effect, but increases the noise margine so that the circuit shows noise immunity up to a certain extent.

Figure 12.19b and c display an OR gate. The Boolean operation is established with FETs in the well-known manner. In combination with the RTD the I-V characteristics of the OR gate are similar to that of the inverter. The RTD offers the advantage of high gain at the switching point, because of the

## 12.3 Digital Circuit Design Based on RTDs

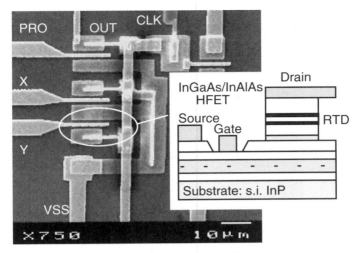

**Fig. 12.15.** Micrograph and cross section of FET-RTD combination, W. Prost, F.-J. Tegude

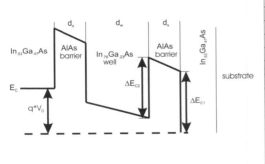

| 100 nm | InGaAs | $N_D = 1{,}5 \times 10^{19} \text{cm}^{-3}$ |
|---|---|---|
| 50 nm | InGaAs | $N_D = 7{,}5 \times 10^{17} \text{cm}^{-3}$ |
| 4 ML | InGaAs | spacer |
| $d_s = 8$ ML | AlAs | barrier |
| 4 ML | InGaAs | smooting |
| $d_w = 8$ ML | InAs | well |
| 4 ML | InGaAs | smooting |
| $d_s = 8$ ML | AlAs | barrier |
| 4 ML | InGaAs | spacer |
| 50 nm | InGaAs | $N_D = 7{,}5 \times 10^{17} \text{cm}^{-3}$ |
| 100 nm | InGaAs | $N_D = 1{,}5 \times 10^{19} \text{cm}^{-3}$ |
| 10 nm | InAlAs | etch stop |
| 300 nm | InGaAs | $N_D = 1{,}5 \times 10^{19} \text{cm}^{-3}$ |
| s.i. -InP:Fe | | substrate |

**Fig. 12.16.** Layer structure of an RTD, courtesy of F.-J. Tegude

negative differential resistance region. Therefore the switching itself is very fast.

### 12.3.3 Dynamic Logic Gates

Taking even more advantage of the NDR region leads to promising concepts. A good example in this regard is the *monostable bistable logic transition element* (MOBILE) [56]. Its plain configuration can be found in Fig. 12.20. In recent years high-speed logic families based on the MOBILE have been proposed for tunneling devices. The MOBILE gate is a pseudo dynamic, clocked

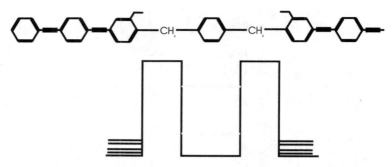

**Fig. 12.17.** Concept of a polymer RTD, J.C. Ellenbogen

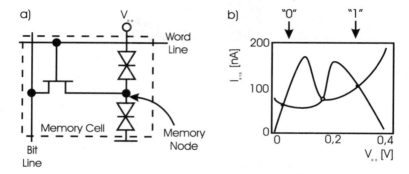

**Fig. 12.18.** Static memory cell based on RTDs

logic circuit and consists of two resonant tunneling diodes that operate in a monostable or bistable state depending on a clocked power-supply voltage. The term pseudo dynamic refers to the special circuit style in the sense that a clock controls the logic transition of the gate, similar to dynamic circuits. But in contrast to dynamic circuits, where the logic state is represented by the electrical charge on a capacitor, MOBILE circuits are in a static, self-stabilizing state due to the inherent bistability of their devices. Consequently, they are more robust against charge leakage, and precharging is unnecessary. The functionality of MOBILE-type gates is specified by embedding a logic input stage into the self-latching RTD pair [57]. This logic input stage can be used to implement a variety of different logic schemes, such as conventional boolean logic, threshold logic, and multivalued logic [58]. Depending on the specific device technology the input stage is composed of HFETs, surface tunneling transistors, RTD-Schottky diode combinations, or RTBTs.

A significant feature of the RTD-HFET is that the peak current is independent of the gate voltage for $V_{GS} > 0.2$ V. Thus, if a logic high voltage is applied to the gate of the RTD-HFET there is no difference between an

12.3 Digital Circuit Design Based on RTDs    201

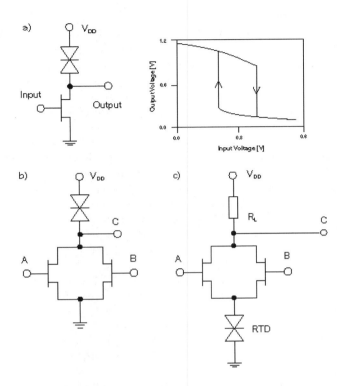

Fig. 12.19. Inverter and logic OR gates based on RTDs.

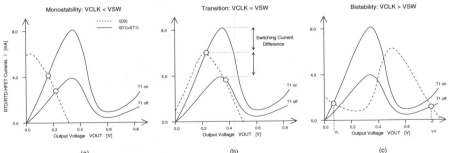

Fig. 12.20. Load-line diagrams of the RTD-HFET MOBILE inverter for increasing clock voltages $V_{CLK}$

RTD-HFET and the two RTDs as it is shown by the equidistant steps of the switching-current difference in Fig. 12.20b.

The logic function of a MOBILE is related to the peak-current difference of load and driver RTD. If the MOBILE input is provided to the gate-source voltage $V_{GS}$ of an HFET this current difference $\Delta I$ is determined by the transfer characteristic $\Delta I = I_D = f(V_{GS})$. Therefore, the logic function becomes dependent on a precise control of the HFET threshold voltage and transconductance. At a sufficiently high input voltage (gate bias $V_{GS} > 0.2\ V$) the drain current is limited by the RTD peak current $I_P$ and is independent of the input voltage. Hence, the transistor is limited to a switching functionality and the precise addition of a current $\Delta I$ to the MOBILE is adjusted by the RTD area ($\Delta I = I_P = \alpha \cdot A_{RTD}$) made from the same layer sequence and with the technology as load and driver RTD ($\alpha = $ const.). This concept enables the manufacture of complex gates with several inputs. Especially the design of multiple-input threshold gates with constant weights determined by the input RTD area only becomes feasible [59]. In addition, the decrease of the drain current at high drain bias results in a lower power consumption in the on-state and a higher driving capability.

To extend the computational capabilities of RTD logic gates the current-controlled switching of the MOBILE in connection with the RTT input stage allows a functional implementation of linear threshold gates (LTGs). The characteristic feature of LTGs is the parallel processing of multiple inputs. A LTG calculates the weighted sum $\chi$ of the digital inputs $x_k, k = 1, \ldots, N$. The weighted sum is converted into a digital output $y$ by comparing $\chi$ with a given threshold value $\Theta$ (Fig. 12.21). If the weights $w_k$ and the threshold value $\Theta$ are selected properly the LTG computes any linear separable Boolean function of the $N$ inputs. The output $y$ of a threshold gate is given by

$$y(\chi) = \text{sign}\,(\chi - \Theta) = \begin{cases} 1 \text{ if } \chi \geq \Theta \\ 0 \text{ if } \chi < \Theta \end{cases} \quad (12.3)$$

$$\chi = \sum_{k=1}^{N} w_k \cdot x_k \quad , \quad x_k = \{0, 1\}$$

$$w_k = \{0, \pm 1, \ldots, \pm w_{max}\}\,,$$

$$\Theta = \{0, \pm 1, \ldots, \pm \Theta_{max}\}\,.$$

Compared to a Boolean logic gate, a threshold gate combines an internal multivalued computation of the weighted sum with digitally encoded input and output states. Actually, this capability of processing multiple input signals enables it to design circuits with bit-level parallelism and reduced complexity. Thus, our approach shows a certain similarity to multivalued logic, but differs in regard to the digital input and output logic states. Although not explicitly mentioned the programmable NAND/NOR gate of the previous subsections is

## 12.3 Digital Circuit Design Based on RTDs

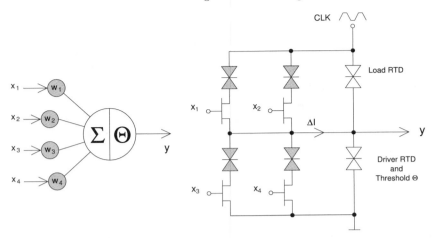

**Fig. 12.21.** Linear threshold gate $y = \text{sign}(x_1 + x_2 - x_3 - x_4 - \Theta)$ and MOBILE circuit

basically a LTG with two negative weighted inputs and a modifiable threshold value.

The circuit in Fig. 12.21 shows an RTD-based threshold gate composed of two serially connected RTDs and four RTTs with two positive-weighted inputs ($x_1, x_2$) and two negative-weighted inputs ($x_3, x_4$). The threshold value is implemented by modifying the RTD anode area of the driver RTD. According to the current-controlled switching the weighted sum $\chi$ of a LTG is then given by the total input current at the output node according to Kirchhoff's current law. If the threshold gate has $M_p$ positive-weighted and $M_n$ negative-weighted inputs, the total input current at the monostable-bistable transition point is

$$\Delta I = \sum_{k=1}^{M_p} w_k I_P(V_{GSk}) - \sum_{k=1}^{M_n} w_k I_P(V_{GSk}) \,. \tag{12.4}$$

Here, $I_P(V_{GSk})$ is the peak current of an RTD with minimum RTD area $A_{min}$ (weight $w_{min} = 1$). The weight factors $w_k$ express the area ratios between the RTDs with weighted inputs and the RTD with minimum area: $A_w = w \cdot A_{min}$. Thus, according to the design methodology of the NAND/NOR gate, the weight factors $w_k$ are equivalent to the RTD design parameters: $|w_k| = \lambda_k$. The inputs $x_k$ of the threshold gate are set by the gate source voltage $V_{GSk}$ and load-line diagrams similar to Fig. 12.20 can be derived. In the proposed circuit the metastable transition is an efficient way to implement the required comparison function because the sign of the input current is equivalent to the sign of the weighted sum. Since the threshold value $\Theta$ is the peak current difference of the driver and load RTD $I_\Theta = \Theta \, I_P$ the internal weighted sum (i.e. the input current) is finally converted into a digital output:

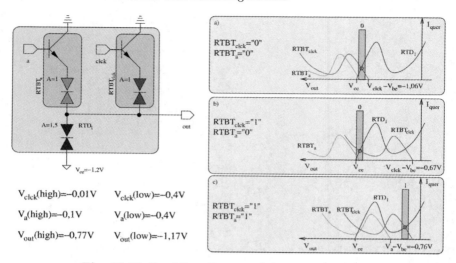

**Fig. 12.22.** Load-line diagram of the basic RTBT gate

$$V_{OUT} = \begin{cases} V_H \text{ for } \Delta I > I_\Theta \\ V_L \text{ for } \Delta I < I_\Theta. \end{cases} \quad (12.5)$$

Figure 12.21 shows a universal RTD gate. The inputs determine the current that flows through the input RTD. When this current is higher than the threshold current $I_{th}$ of the MOBILE gate the gate switches to the logic high-level. The logic function of the gate can be defined by the tuning of the individual input currents. Although this gate appears to be a very elegant and universal implementation, it suffers from the disadvantage to be nonimmunized to signal fluctuations. In terms of digital VLSI design the immunity against any amplitude characteristics is of fundamental importance. This claim seems to hold even when it comes to nanoelectronics.

## 12.4 Digital Circuit Design Based on the RTBT

### 12.4.1 RTBT MOBILE

The RTBT MOBILE is based on the same principle as the HFET-RTD MOBILE introduced in the previous section. Two serially connected resonant tunneling structures offer a monostable-bistable element that can be controlled by a single clock signal [60].

Figure 12.22 shows the monostable-bistable transition of an RTBT input stage. The areas of the RTDs are chosen so that the peak current of the load $RTD_1$ is larger than the current of $RTBT_{clck}$, but smaller than the current sum of $RTBT_{clck}$ and $RTBT_a$. From this it follows that the gate switches to the logic low voltage if the input $RTBT_a$ is off. If $RTBT_a$ is on, the current sum of $RTBT_{clck}$ and $RTBT_a$ exceeds the current of the load $RTD_1$ and the gate switches to the logic high voltage.

The threshold is determined by the peak current of $RTD_1$ while the input weight is represented by the peak current of $RTBT_a$. An input weight of $w = 1$, for example, is characterized by the minimum cross section of the resonant tunneling structure in the emitter branch of the HBT. This scales the peak current to its minimum and is symbolized by $A = 1$. The gate switches from the monostable state into the bistable state if the clock voltage $V_{clck}$ drives $RTD_1$ over its peak voltage. The bistable states represent the two logic levels 0 (low) and 1 (high). At the switching point the gate follows the upper or lower gate characteristics depending on the input voltage $V_a$. The peak voltage $V_P$ and the clock voltage $V_{clck}$ is 0.18 $V$, and $-0.1$ $V$ respectively. This leads to a voltage swing of $\Delta V = 0.4$ $V$. During the active clock phase the voltage drop $V_{clck} - V_{be} - V_{ee} = 0.53$ $V$ across the two resonant tunneling devices lies between $2V_p = 0.36$ $V$ and $3V_p = 0.54$ $V$ and drives the MOBILE from the monostable into the bistable state.

Comparing the RTBT MOBILE with the HFET MOBILE concept of Chen and its extension by Pacha it uses the monolithically grown RTBT input stage for both the signal input and the clock input. As the MOBILE is a clocked and current-controlled gate with integrated latch function it is well suited to the current-controlled RTBT. Furthermore, the lack of a reliable enhancement-type HFET is not an issue for the RTBT because of its built-in potential that depends on the base and emitter bandgap.

### 12.4.2 RTBT Threshold Gate

A linear threshold gate calculates the weighted sum $\chi$ of the digital inputs $x_k, k = 1, .., N$. By comparing $\chi$ with a given threshold value $\Theta$, the weighted sum is converted into a digital output (Fig. 12.21).

Figure 12.23 shows the schematic view and the micrograph of the RTBT threshold stage. The weight of each input is determined by the peak current of the corresponding resonant tunneling diode (RTD). Since the heterojunction bipolar transistor (HBT) is a current-driven device we use an emitter-follower configuration. This keeps the input impedance high and the transistor does not go into saturation. To realize negative weights it should be reasonable to add RTBT input stages into the pull-down network (PDN), but this is problematic for two reasons: First, the RTBTs in the PDN need a dc offset of at least $V_{BE(on)}$ to maintain I-O compatibility. This additional logic level would require a level shifter. Secondly, the RTBTs of the extra input stages tend to enter saturation which should be omitted for performance reasons. Instead we use a modified differential output buffer. As we will see later, the output buffer preserves the threshold gate from the drawbacks of the input stage in the PDN, while it enables negative weights.

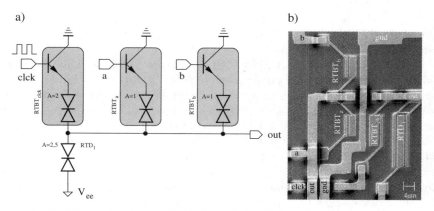

**Fig. 12.23.** Schematic view and chipfoto of the RTBT threshold gate, W. Prost, F.J. Tegude.

## 12.4.3 RTBT Multiplexer

The steady movement towards broadband integrated services digital network implies an expansion of transmission capacity to a level beyond multi-terabit per second. High-speed time-division multiplexers are key components in broadband communication systems. In this connection RTBT based MOBILE circuits help to reduce the component counts per logic function. An increase of the logic functionality of the single device keeps the logic depth low and reduces its complexity. Figure 12.24 shows the schematic view of the suggested 2:1 RTBT multiplexer IC. Each input stage is based on the RTBT MOBILE introduced before in Fig. 12.25 and behaves as a logic AND. The two inputs $data_a$ and $data_b$ are used as signal inputs whereas the inputs $clck$ and $\overline{clck}$ select the individual channel of the multiplexer. Applying a logic low level to one of the clock inputs will lock the individual input stage in a logic low level, meanwhile the output of the other inverter is ruled by its data input.

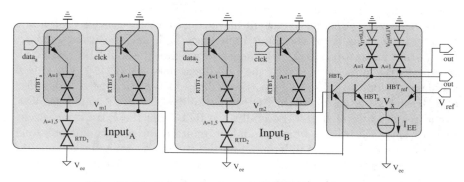

**Fig. 12.24.** Schematic view of the RTBT multiplexer

## 12.4 Digital Circuit Design Based on the RTBT

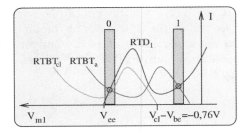

**Fig. 12.25.** Load-line diagram of a single input channel

The output of the MUX is established by a modified ECL buffer. The two parallel transistors $HBT_a$ and $HBT_b$ can share the same branch of the ECL buffer, since they are never activated simultaneously. Considering the design methodology the RTBT-MUX consists of simple input stages being used as basic building blocks. Hence, a regular design structure is maintained and the RTBT-MUX circuit is characterized by a low output impedance and a high input impedance.

Multiplexers for the next generation of digital receivers, ATM switches, and buffers depend on such highly integrated functional blocks.

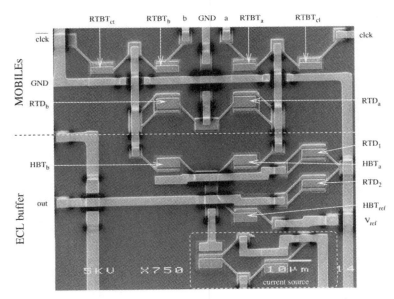

**Fig. 12.26.** Micrograph of the RTBT multiplexer (A≈ 125 × 130 $\mu m^2$), F.J. Tegude

The micrograph of the RTBT multiplexer can be seen in Fig. 12.26. The symmetrical format as well as the separation of the input and output stage can also be traced in the implementation of the design. Due to the vertical structure of the monolithically grown RTBT it can not be distinguished in the top view from the HBT.

## 12.5 Summary

Tunneling devices are controlled by the applied voltage. There are two effects: First, the electric field distorts the potential barrier so it becomes thinner and the tunneling current larger, secondly, the potential drop moves the allowed bands so the current changes or stops. The combination with a transistor overcomes the disadvantage of the only two-terminal device.

From the present state of technology, circuits with tunneling elements are the first step into nanoelectronics. At the moment the special technologies do not offer the chance to realize lateral nanostructures since this technology is not so well developed as the silicon technology because of economic reasons. Therefore the main applications of tunneling elements are fast logic circuitries and fast memories.

# 13
# Single-Electron Transistor (SET)

One of the most exciting challenges of microelectronics is, of course, the vision of a switching device that can control single electrons. The idea of the single-electron transistor is based on the so-called *Coulomb blockade*. This effect only appears within very small device dimensions and is due to the quantization of charge [61].

## 13.1 Principle of the Single-Electron Transistor

### 13.1.1 The Coulomb Blockade

To explain the Coulomb blockade we take a closer look in Fig. 13.1 at the composition of a tunneling element and a capacitor. The capacitor is located on the right side of the tunneling element and has a very small capacitance due to its very small dimensions. The energy level of electrons within the capacitor can be adjusted by an external gate voltage $V_G$.

Electrons can tunnel from the right side of the barrier to its left side as long as the energy level on the right side remains lower than on the other side. If this is not the case, the electrons get blocked by the Coulomb blockade, since the charge of a single electron would cause an increase of the energy level. This remarkable increase is due to the very small capacitance of the capacitor. Electrons can only tunnel if the energy balance is positive after tunneling. This effect can be explained within the boundaries of classical physics by Coulomb's law, however, it assumes the quantization of the electrical charge. The tunneling element is needed for a galvanic isolation, but it nevertheless offers a voltage-controlled tunneling probability.

The following discourse derives the most important equations of the single-electron transistor. The total energy E depends on the amount of charge on the capacitor and on the electricical potential V. The energy of a capacitor C that is charged with n electrons is equivalent to:

# 13 Single-Electron Transistor (SET)

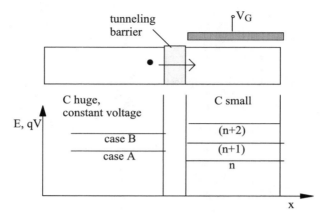

**Fig. 13.1.** Schematic cross section of the composition 'tunneling element and capacitor' as well as its energy diagram, case A: the electron cannot tunnel since its potential energy would be higher afterwards, case B: the electron tunnels because the electron's energy balance is positive after tunneling, however a further electron gets blocked again by the Coulomb blockade. The energy shift from A to B is controlled by the gate voltage $V_G$ (13.3)

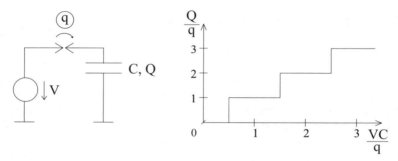

**Fig. 13.2.** Equivalent circuit diagram of the composition 'tunneling element and the capacitor' with $Q = VC$ as well as their Q-V characteristics for the normalized voltage

$$E_n = \frac{1}{2}\frac{n^2 q^2}{C}. \tag{13.1}$$

The energy difference on the capacitor C for the two charge situations of n and n+1 electrons corresponds to:

$$E_{n+1} - E_n = \frac{q^2}{2C}(2n + 1). \tag{13.2}$$

An electron can only pass through the barrier via tunneling if this energy difference is less than or equal to the energy of an electron. The energy of an

## 13.1 Principle of the Single-Electron Transistor

electron can be expressed as:
$$E = qV. \tag{13.3}$$
This implies that the voltage V should not exceed the following value:
$$V = \frac{q(2n+1)}{2C}, \quad n = 0, 1, 2, 3, \dots . \tag{13.4}$$
For the normalized voltage the location of the steps is equivalent to:
$$\frac{VC}{q} = \frac{1}{2}(2n+1). \tag{13.5}$$

According to $n = \frac{Q}{q} = 0, 1, 2, 3, \dots$ Fig. 13.2 reveals the stair function: Each step originates from the tunneling of a further electron onto the capacitor. If the capacitor is scaled down towards a quantum dot, distinct energy levels arise due to the de Broglie waves of the electron. This effect of quantum confinement goes beyond the Coulomb blockade.

### 13.1.2 Performance of the Single-Electron Transistor

The SET consists out of two tunneling barriers and a small island (Fig. 13.3). The gate voltage controls, via the Coulomb blockade, the current that flows from the source to the drain contact. In contrast to devices such as RTDs and FETs the switching behavior of the SET depends on single elementary charges. For an explanation we take a closer look at the equivalent circuit diagram in Fig. 13.4.

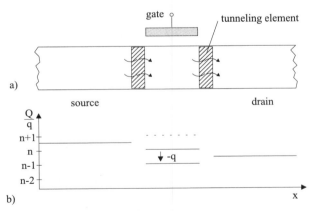

**Fig. 13.3.** Setup of the SET and its corresponding energy diagram

Starting from the above-derived relation we can state the voltage condition that has to be met for tunneling if $V_{GS} = 0$:

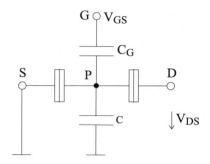

**Fig. 13.4.** Equivalent circuit diagram, the island P is located in between the tunneling elements

$$V_{PT} = \frac{q}{2(C + C_G)}. \tag{13.6}$$

Since the island owns has as much voltage as the drain contact the current $I_{DS}$ starts to flow at the voltage:

$$V_{DS} = \frac{q}{C + C_G}. \tag{13.7}$$

Figure 13.5 refers to this threshold voltage as case 1. The corresponding characteristic is symmetrical about the zero point. The current vanishes in the so-called *Coulomb gap*. The voltage $V_{PS}$ between the island P and the source contact S depends on $V_G$:

$$V_{PS} = \frac{C_G}{C + C_G} V_G. \tag{13.8}$$

However, a current can only flow if this voltage corresponds to the above-derived threshold voltage. Under these circumstances, it follows from these two equations:

$$V_G = \frac{q}{2C_G}. \tag{13.9}$$

Figure 13.5 refers to this as case 2. For this gate voltage the Coulomb gap disappears. Figure 13.5 also depicts the threshold voltage $V_T$ as a function of the gate voltage $V_G$. The interplay of charge packets results in a symmetrical alignment of periodic and even functions.

However, for the complete description of the overall transistor performance there are more parameters to be considered:

1. Charge that is located in the surroundings of the transistor has an impact on its threshold voltage and may cause an uncontrolled drift of the transistor's threshold voltage. At present this is considered to be a serious

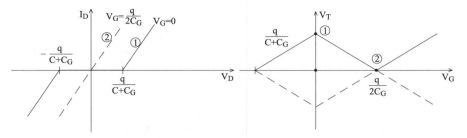

**Fig. 13.5.** Characteristics of the drain current versus the drain voltage and of the threshold voltage characteristics versus the gate voltage for a SET

problem, since the background charge is, in many cases, time dependent. It appears as some kind of noise. A background charge of $Q_0$ results in a threshold voltage drift of:

$$\Delta V = \frac{Q_0}{C + C_G}. \tag{13.10}$$

2. As a further restriction the thermal energy has to be considerable lower than the electrostatic energy $E_C$.

$$E_C = \frac{q^2}{C} \gg kT. \tag{13.11}$$

This constraint can already be met with moderate-scaled devices if the circuit is cooled. If the capacitances are scaled down in a more aggressive manner the energy steps will increase in such a way that the circuit can be operated at higher temperatures. Consequently, if the capacitance of the island can be scaled down to a capacity of roughly $10^{-18}\,F$ roomtemperature operation will become feasible.

3. The switching behavior depends on the time constant of the tunneling process and is dominated by the tunneling resistance and the overall capacitance:

$$\tau_T = R_T C. \tag{13.12}$$

As a further constraint the electrons should cross the barrier because of the tunneling process and should not interfere with the uncertainty principle:

$$E_c \tau_T \gg h. \tag{13.13}$$

The energy $E_c$ has to be considerable higher than the thermal energy as explained in Chap. 4:

$$E_c = \frac{1}{2}CV^2 \gg kT. \tag{13.14}$$

Therefore, the tunneling resistance is restricted to:

$$R_T \gg 2\frac{h}{q^2}. \qquad (13.15)$$

This quantity is called the von-Klitzing resistance and evaluates to:

$$R_K = \frac{h}{q^2} = 25.8\,k\Omega. \qquad (13.16)$$

Obviously, SET circuits show a higher impedance than conventional CMOS circuits. This is not a critical issue since the accompanying capacitances are relatively small.
The above mentioned time constant can be evaluated at $T = 6\,K$:

$$\tau_T \geq \frac{h}{kT} \approx 10^{-11}\,s. \qquad (13.17)$$

For a capacitance of $C = 2 \times 10^{-16}\,F$ the time constant evaluates to:

$$\tau_T = 2R_K C = 2 \times 10^{-11}\,s. \qquad (13.18)$$

Both cases show almost the same time constant. For application purposes it will be about $1\,ns$, which is quite small for such a high transistor impedance.

The presented model of SET is an idealistic one, real circuit simulation needs a more sophisticated model [62].

### 13.1.3 Technology

At first glance it seems to be difficult to fabricate SETs consisting of two tunneling elements and a small island. The following three concepts will reveal the feasibility of the SET technology. The technological challenge is to find the appropriate way to bring one of these concepts into mass production [63].

The first concepts uses a batch process to coat small metal balls with an insulator. Such a ball can be placed on top of an interrupted conductor. The gate might be integrated into the substrate (Fig. 13.6a). The objective of such concepts is to achieve some kind of self-organization on the level of atoms or molecular clusters.

Another concept realizes the appropriate tunneling elements via different deposition angles (Fig. 13.6b). First, Al is deposited from the right side so that its oxide can directly be used as a tunneling barrier. In a second process step Al is applied from the left side as depicted in the schematic view of Fig. 13.6b. The two applied Al layers overlap and form the tunneling element.

In silicon the SOI technique can also be used to implement SETs. As Fig. 13.7 reveals the essential elements such as tunneling barriers and islands can be realized within a very thin Si layer. The use of the substrate as gate electrode is feasible, but is not suitable for fast circuit operations. This concept might be improved by an additional silicon layer that includes the gate electrode.

13.1 Principle of the Single-Electron Transistor      215

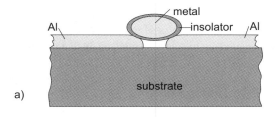

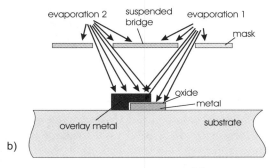

**Fig. 13.6.** Different concepts for SET fabrication: (a) metal balls that are coated by an insulator, (b) different deposition angles result in two tunneling elements and an island

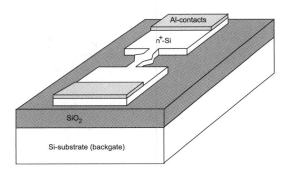

**Fig. 13.7.** SOI technique in silicon: The two thin bridges connect the island to its surroundings, the substrate forms the gate electrode, which is not too convenient for circuit design

This example shows distinctly that it is very important to find an appropriate structure for the SET implementation. The contacts have to have small parasitic capacitance, meanwhile the tunneling contacts have to be reliable.

## 13.2 SET Circuit Design

The following examples explain the principles of SET circuit design. However, their practical operation is still a long way off. The SET circuit design fashion is strongly motivated by typical CMOS design techniques as well as QCA design techniques. From the present point of view SET circuits are very promising candidates for memory applications, artifical neural networks, and low-power applications [64].

### 13.2.1 Wiring and Drivers

The charge of a single electron can not drive a metallic wire. Consequently, the charging of a long wire via a SET would take a long time. A possible solution might be a shift register that acts as a wire. Within such a shift register the data is represented by single electrons that are shifted from one node to the other (Fig. 13.8). For such a configuration the (dis)charging of a long wire is not a issue. However, the shift register needs a clock supply for a fast data exchange, which consumes electrical energy. A similar concept has been presented for the QCA circuit design in Chap. 10.

Two clocked SETs can be operated as a current driver (Fig. 13.9). They act as an electron pump: If the gate voltage is altered with frequency $f$ within each clock cycle one electron can pass from the left electrode via the two tunneling elements to the right side. Therefore, the current $I$ is proportional to the frequency f:

$$I = qf. \qquad (13.19)$$

An atomic clock might control the frequency f so that a very precise control of the current is feasible. This is a very interesting issue for measurement standards.

For circuit designers the driving capabilities are more interesting: The characteristic impedances of the circuits are relatively high, however, the capacitances are small and the pumped current increases with the frequency. Assuming resistances of $100\,k\Omega$ and capacitances of $10^{-16}\,F$ the switching time is about $10\,ps$. Using a clock period of $10\,ps$ it is possible to pump 1 to $10^2$ electrons onto the output capacitor within $1\,ns$. This number of electrons can already drive a CMOS interface circuit, in particular if the outputs can be driven in a parallel way.

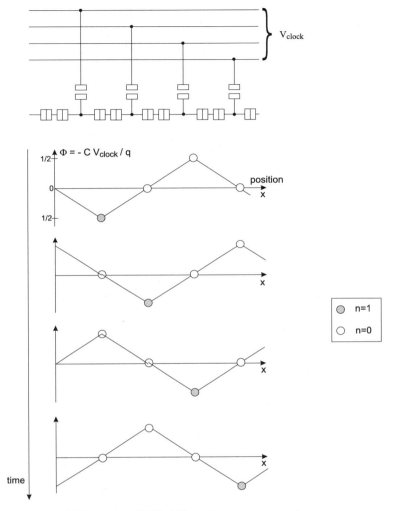

**Fig. 13.8.** SET shift register acts as a wire

### 13.2.2 Logic and Memory Circuits

According to Fig. 13.5 the transistor toggles with the gate voltage between a conducting and an insulating state [65]. Two serial-connected SETs can exploit this effect in such a way that they operate as an inverter (Fig. 13.10). Depending on the input signal only one of the two SETs is in the conducting state. In addition, Fig. 13.10 illustrates a programmable logic gate. Via the control input the gate operates as a logic NOR or NAND gate.

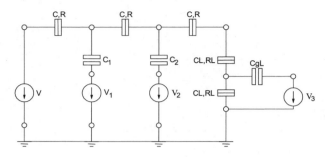

**Fig. 13.9.** Electron pump: Its output produces an active current

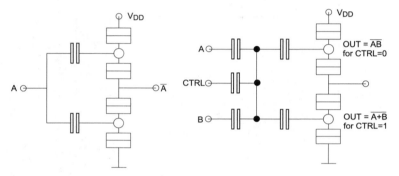

**Fig. 13.10.** Inverter and programmable logic gate

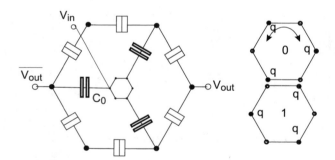

**Fig. 13.11.** Six SETs form a ring-shaped memory cell: The stored information depends on the location of 3 electrons

In principle, it is possible to build a memory cell out of two inverters. An even smarter solution is a ring-shaped memory cell that only shifts electrons between the islands (Fig. 13.11). Similar to a dynamic memory cell (DRAM) the information can also be stored as charge on a capacitance [66].

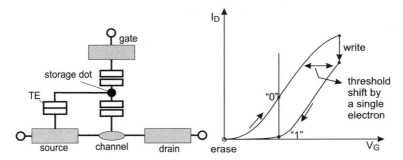

**Fig. 13.12.** Memory cell with SET and MOS transistor

An interesting version of such a memory cell can be seen in Fig. 13.12. The SET is integrated into the MOST. For high gate voltages the SET pushes a further electron on its island. This enlarges the threshold voltage of the MOS transistor. For lower gate voltages the SETs island gets discharged again.

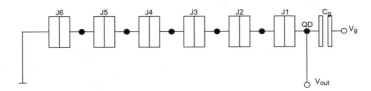

**Fig. 13.13.** SET memory chain stores several electrons

The SET memory chain in Fig. 13.13 can be used to store several electrons on a single memory node. If the voltage $V_G$ is tuned in an appropriate way the electrons move to the left side of the chain. On each island a single electron gets trapped by means of the Coulomb blockade. If $n$ is the number of tunneling elements of the chain structure $(n-1)$ electrons can be captured by this method. The electrons move towards the capacitance $C_G$ if the voltage $V_G$ is readjusted. After the discharge process the voltage $V_{out}$ corresponds to the number of electrons that have been previously in the chain. This memory chain might be used for a multivalued data storage, but at present this structure is only used to store binary data because of the faint signals.

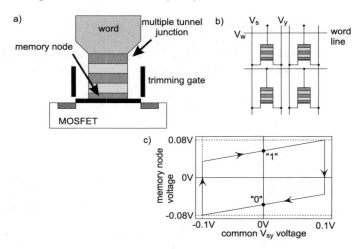

**Fig. 13.14.** Realization of a SET memory mash

A realization of such a memory structure is illustrated in Fig. 13.14. Each tunneling element consists of silicon particles that are composed in a polycrystalline stack. Every stack can store about 100 electrons, which results in a robust signal for the MOS transistor.

### 13.2.3 SET Adder as an Example of a Distributed Circuit

Figure 13.15 depicts the circuit diagram of a SET adder. The distributed circuit has been designed in a modular and repeating manner, which is of particular importance for nanoelectronic circuits. The input data is fed into the adder by means of capacitive coupling. Four individual clocks control the processing of the data. Figure 13.16 takes a closer look at a single adder cell.

The nodes A and B introduce the input data. If an electron is located in node A it is shifted to node B during the first clock cycle. This shift is only possible if node B is not occupied. In this case the electron moves back to node A during the second clock cycle and switches the lower SET into the conducting state. As a consequence, the electron in node B disappears. The third and fourth clock cycle are reserved for the carry. This example shows that complex SET circuits without long wiring distances are feasible.

## 13.3 Comparison Between FET and SET Circuit Designs

The comparison between FET and SET as well as their circuit designs is a very interesting issue within the power-delay diagram. Although the power-delay diagram only offers a simplified view of both concepts the fields for potential

## 13.3 Comparison Between FET and SET Circuit Designs

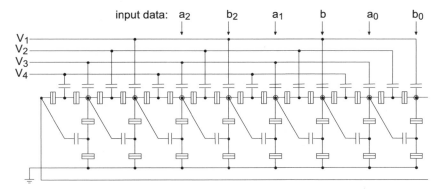

**Fig. 13.15.** SET adder as distributed circuits (T. Ramcke, W. Rösner, and L. Risch)

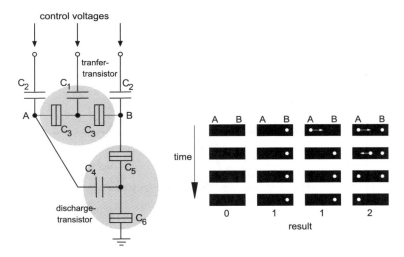

**Fig. 13.16.** A single adder cell of the SET adder of Fig. 13.15

applications can be derived. Figure 13.17 shows the normalized switching energy $W_S/W_{S0}$ as a function of the scaling factor $\alpha$. $W_{S0}$ refers to the switching energy at $\alpha = 1$.

The switching energy of a FET decreases as the device dimensions decrease, because the number of electrons in the channel as well as its capacitance scale with a factor of $\alpha^2$. This scaling procedure can only be continued until the last electron is left in the channel. This is the point where the FET model turns into the SET model. If the scaling of the physical device dimensions is continued beyond this point, the charge of the last electron obviously can not be scaled. Therefore, the switching energy "scales" from this point on

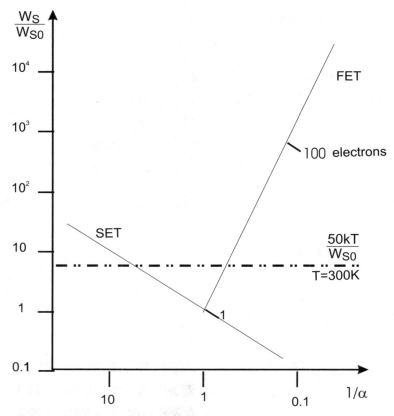

**Fig. 13.17.** The power-delay diagram reveals the impacts of scaling on the SET and FET

with $\alpha^{-1}$, which actually means that the switching energy starts to increase for smaller dimensions.

For a specific case the FET curve in Fig. 13.17 is labeled with the number of electrons that are involved in each switching event. The presently used CMOS technology handles roughly 10 000 electrons during a single switching event. The thermal limit at $300\,K$ is depicted in Fig. 13.17 for comparison purposes and is defined as $W_S = 50kT$. As denoted in Chap. 8 this is the minimum energy that has to be applied to avoid thermal hazards. To operate these circuits at $300\,K$ the SET device dimensions have to be very small, while the FET device dimensions have to be relatively large. In between these two extremes a save operation is only possible via cooling.

This consideration is of fundamental relevance for nanoelectronics: For relatively large device dimensions the energy stored in the capacitor can only cope with the above-mentioned constraints if the thermal energy is reduced via cooling. For smaller device dimensions the switching energy becomes higher than the thermal energy and no additional cooling is necessary. The increase of the stored energy for scaled devices is typical of quantum-effect devices such as superconductor switching elements. Their quantized quantity is the magnetic flux and will be described in the next chapter.

## 13.4 Summary

Single electron transistors offer new and interesting circuit-design concepts since they can be controlled by a single elementary charge. For small device dimensions the switching energy is so high that the SETs can be operated at moderate temperatures. The impact of disturbing effects such as background charge still remain a challenging issue. Besides further technologically oriented improvements the investigation of suitable logic families, architectures, and the development of a design framework are of fundamental relevance.

# 14
# Nanoelectronics with Superconducting Devices

Superconductivity is a fascinating effect from which we assume that devices with ideal characteristics can be built. It turned out that the very low resistances of a switched superconductor have a negative impact on its transient behavior. Superconductivity is not a good candidate for further microminiaturizing except for some highly specialized applications due to the increase in switching energy with decreasing feature sizes. The reason for this effect is based on the fact that a flux quantum is more efficient than the electronic charge. Nevertheless, the research on superconducting electronics is valuable and important, especially with regard to nanoelectronics.

## 14.1 Basics

### 14.1.1 Macroscopic Characteristics

The most popular characteristic of a superconductor is its vanishing resistance below a certain temperature, the transition temperature $T_C$. With the vanishing resistance there occurs the *Meißner-Ochsenfeld* effect: A magnetic field is displaced out of the superconductor by surface currents, Fig. 14.1. In this case a superconductor and an infinitely good conducting material differ. The latter shows no change in the field characteristics within the conductor.

A very important diagram for the characterization of a superconductor is the $H-T$ diagram in Fig. 14.2, which depicts the transition from the normal conducting to the superconducting state. Applying a sufficiently large magnetic field $H$ to the superconductor switches it from the superconducting to the normal conducting state. This happens even far below the transition temperature. For the condition $T = 0\,K$ this field strength is called the critical field strength $H_C$. A current inside the superconductor results in a magnetic field, which in turn is able to change the state of the superconductor.

The transition temperatures and the critical field strengths of some superconductors as well as selected boiling temperatures and coolants are summarized in Table 14.1.

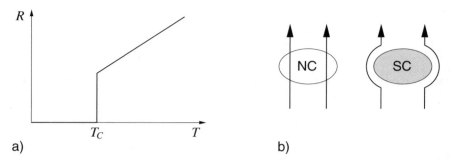

**Fig. 14.1.** Characteristics of an ideal superconductor: Resistance in relation to the temperature (a), displacement of the magnetic field (b)

**Table 14.1.** Transition temperatures and boiling points of selected superconductors and cooling agents

| Superconductor 1st kind | | Superconductor 2nd kind | | High-$T_C$ superconductor | | Boiling points | |
|---|---|---|---|---|---|---|---|
| $Al$ | $1.2\,K$ | $Nb$ | $9.2\,K$ | $YBa_2Cu_3O_7$ | $92\,K$ | $He$ | $4.2\,K$ |
| $In$ | $3.4\,K$ | $NbN$ | $16\,K$ | | | $N_2$ | $77\,K$ |
| $Sn$ | $3.7\,K$ | $Nb_3Ge$ | $23\,K$ | | | | |
| $Pb$ | $7.2\,K$ | | | | | | |

We distinguish between superconductors of the 1st kind, which show widely ideal characteristics and are applied in electronics, and superconductors of the 2nd kind with higher transition temperatures applied in heavy current engineering. The latter does not displace the magnetic field completely. High-temperature superconductors, first discovered in 1983, are promising candidates for a broad range of applications but a comprehensive research activity is missing. New materials with higher transition temperatures are discovered continually. Materials with transition temperatures of $T = 135\,K$ exist. At

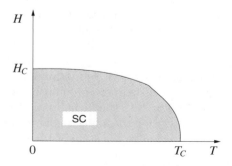

**Fig. 14.2.** State diagram of a superconductor

present superconductors at room temperature are under investigation. High-temperature superconductors are based on ceramic material that are difficult to manage in electronics.

Another important aspect for the application of superconductors are the boiling points of the cooling agents, for example liquid helium ($T = 4.2\,K$) and liquid nitrogen ($T = 77\,K$). High-temperature superconductors can be cooled in a less expensive way with liquid nitrogen. Today achieving low temperatures is not a problem at all. Refrigeration units of different sizes are available at reasonable prices but consume a large amount of space and energy. Most of these units contain some mechanical devices that are more susceptible to faults and are therefore more unreliable than integrated circuits.

### 14.1.2 The Macroscopic Model

In the 1960s the first microscopic models for superconductivity were established. According to the *BCS* theory electron pairs interact and form so-called *Cooper pairs*. The energy of a Cooper pair is smaller than those of two electrons therefore, similar to a semiconductor, a bandgap of $2\Delta E$ is formed, see Fig. 14.3. The bandgap is $3.5\,k_B\,T_C$ and increases with the transition temperature $T_C$ of the superconductor and with decreasing working temperature. Additionally, the model shows that next to the superconducting electrons nonsuperconducting electrons are present.

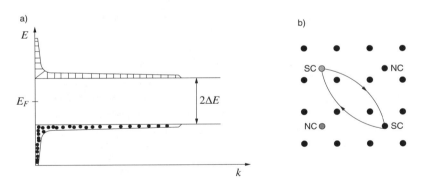

**Fig. 14.3.** Band diagram of a superconductor with a bandgap of $2\Delta E$ at the Fermi level (a). Concurrent existence of electrons NC and Cooper pairs SC (b)

If we apply a direct current to a superconductor only the Cooper pairs cause the current flow, whereas the normal conducting electrons do not move with the applied field, see Fig. 14.3b. Several collisions with the lattice slow down their movement before they are short-circuited by superconducting electrons. In the case of superconducting electrons both electrons of a pair must scatter at the lattice simultaneously with the same amount of energy loss.

This is very unlikely, therefore the resistance is zero. If we increase the current flow the kinetic energy of the Cooper pairs is also increased. If this energy exceeds the bonding energy $2\Delta E$ the Cooper pair splits resulting in a loss of superconductivity. Here the critical current intensity $I_C$ is reached.

If we apply an alternating current the behavior of superconductors is described with the model of *London*. An alternating field accelerates both the Cooper pairs and the normal conducting electrons within the superconductor, see Fig. 14.4. Collisions with the lattice cause a slow down of the normal conducting electrons and increases the energy of the lattice. Taking this into account the equivalent electric circuit in Fig. 14.4b has two branches, one with only an inductance to represent the superconducting electrons and another with an additional ohmic resistor for normal conducting electrons.

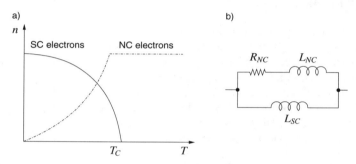

**Fig. 14.4.** Behavior of superconductors when applied to alternating current: Density of both types of electrons in relation to the temperature (a), equivalent electrical circuit (b)

The quality factor of a resonator built of superconducting devices is not perfect because both types of electrons are responsible for the effective current flow when operated with alternating current. Nevertheless, the quality factors of superconducting resonators are better than those of nonsuperconducting resonators, which makes them valuable in practice. This characteristic has also a beneficial effect for superconducting interconnection lines on integrated circuits.

## 14.2 Superconducting Switching Devices

The following deals with different superconducting devices from past to present to give a first impression of the development in this area.

### 14.2.1 Cryotron

A simple device with present minor relevance is the *cryotron* as depicted in Fig. 14.5a.

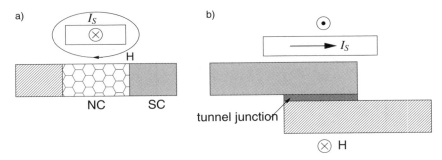

**Fig. 14.5.** Structure and mode of operation of a cryotron (a) and a Josephson device (b)

A control line with a high transition temperature crosses another line. At the junction the line can be switched between the superconducting and normal conducting state by applying a control current. Usually the lines are made of lead and tin and are operated just below the transition temperature of tin. The permanently superconducting control line (lead) can toggle the underlying line (tin) between the superconducting and normal conducting state. Therefore, a cryotron equals a switch that controls the resistance between zero and a limited (small) resistance.

The drawbacks of the cryotron are its low performance due to a phase transition, the difficulty to microminiaturize the device, and the quite small resistance in the normal conducting state.

### 14.2.2 The Josephson Tunneling Device

The drawbacks of the cryotron are mostly avoided by the Josephson device. In relation to Fig. 14.5b an additional tunneling device at the junction of both lines is applied. A 3 $nm$ thick tin-oxide layer as a tunnel barrier is placed on the wire made of tin. Similar to a semiconductor tunnel diode the voltage-current characteristic depends on the bandgap of the structure.

The Cooper pairs tunnel through the barrier, which is equal to a superconducting junction. Therefore no voltage drops at the tunneling device, see Fig. 14.6. When reaching the critical current the bias point moves to the right and the voltage drop across the tunneling device is proportional to the band gap $V_S = 2\Delta E/q$. The voltage remains constant even if the current decreases. The device returns to the superconducting state if the current is reduced to zero.

If tunneling occurs within the right side of the characteristic the Cooper pairs deliver energy by means of photons, depending on the width of the bandgap. In this area, high-frequency oscillations within the Josephson device occur. Their frequency is proportional to the voltage drop across the tunneling device, (14.1).

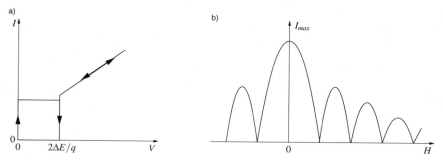

**Fig. 14.6.** (a) V-I characteristic of a Josephson device, (b) state-chart of the device in presence of a magnetic field (b)

$$f = \frac{2q}{h} V_S. \qquad (14.1)$$

Such devices can be used as oscillating circuits and are interesting in the field of neural networks with chaotic behavior. The voltage can be precisely controlled by the frequency, hence we can use this device as a voltage standard.

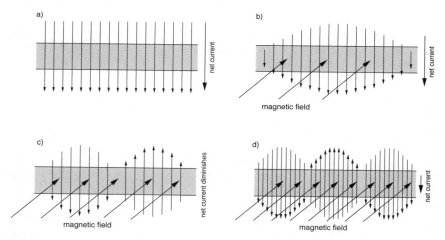

**Fig. 14.7.** Current distribution inside a Josephson device with regard to a magnetic field controlled by a current. The magnetic field increases from (a) to (d)

If a magnetic field is applied to the tunneling device a transition to a high-resistance state occurs and the current disappears. First, the current density at the tunnel junction is uniformly distributed. When increasing the magnetic field the tunnel junction reduces the current. This effect is due to the interaction of tunneling Cooper pairs with the flux quanta. If the current distribution comprises a full wavelength, the net current density is zero and the current through the tunnel junction disappears. In this case the magnetic flux through the tunnel junction is equivalent to one flux quantum. If the magnetic field increases, additional flux quanta can penetrate into the tunnel junction, which results in further switching operations in accordance with the number of flux quantum, see Fig. 14.7. The maximum peaks become smaller because more wavelengths determine the net current. The minima of the current occur at multiples of the magnetic flux quanta. These flux quanta represent the data signals within the *SFQ*-logic. Due to the absence of a phase transition of the material the switching time is very short (some picoseconds). Additionally, the resistance is relatively high. The last-mentioned are advantageous compared to the cryotron, which results in the displacement of cryotrons in electronics.

## 14.3 Elementary Circuits

In the following several basic circuits with superconducting electronics will be presented.

### 14.3.1 Memory Cell

Due to the permanent current flow in a superconducting ring the implementation of memory cells seems to be very attractive. A memory cell can be composed of a superconducting ring where the states are represented by a current flow or not. A circuit technique can use a Josephson device or a cryotron for composing a switch as shown in Fig. 14.8.

Mode of operation: If a current $I_0$ is applied to the ring the current is shared between the two branches in the ratio of the two inductances $L_1$ and $L_2$, (14.2).

$$I_1 = \frac{L_2}{L_1 + L_2} I_0 \quad , \quad I_2 = \frac{L_1}{L_1 + L_2} I_0. \tag{14.2}$$

The necessary condition for this operation is that the magnetic flux inside the ring is zero.

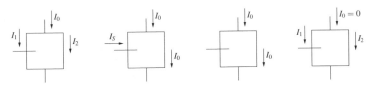

**Fig. 14.8.** Memory cell: If the information is stored the current flow occurs only in the right branch. Not until the externally applied current is switched off does a ring current begin to flow, $I_1 = -I_2$

If the cryotron is switched to the normal conducting state the total current flows through the other branch: $I_2 = I_0$. When switching off the cryotron the condition is maintained. If the current $I_0$ is turned off the current distribution in the ring changes in such a way that the magnetic flux is kept constant inside the ring.

$$I_1 = -I_2 = \frac{-L2}{L_1 + L_2} I_0. \tag{14.3}$$

The information can be sensed destructively by switching on the cryotron. The current and the magnetic flux decay and a voltage, which can be measured externally, is induced. The information is nondestructively readable if the ring current is sensed by an additional cryotron in a special read wire.

### 14.3.2 Associative or Content-Addressable Memory

Storing of information in an associative memory cell is equal to the above-mentioned case. An additional wire composed as a ring wire acts as a switch for controlling the information processing. If a current flows in this branch an ohmic resistance occurs inside the switch, with no current flow the switch remains superconducting. To store an association the corresponding data signals are externally applied to both rings. If the applied signals match with the stored signals the current inside the branch is zero and the switch remains superconducting. If the signals match within a whole word the entire wire is in the superconducting state. Superconductivity is advantageous when sensing the information. Among a great number of data lines the current is able to detect the superconducting line and therefore the associating word. The zero resistance is an outstanding characteristic that can yield a large associative memory.

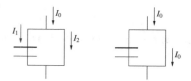

**Fig. 14.9.** Memory cell: Only a current flow inside the left branch activates the switch in the crossing wire (thin line)

The switching performance shows a fundamental difference from those of MOS circuits. One of the main interests in the design of transistor circuits is a fast recharge of the internal capacitances, i.e. to minimize the resistance $R$ in the time constant $\tau = RC$. In terms of superconducting electronics the current has to decay as fast as possible, as shown for the memory cell mentioned above. In this case the time constant is $\tau = L/R$, i.e. the larger the resistance the smaller the time constant. This condition can not easily be fulfilled with superconducting electronics.

The vanishing resistance ($R \to 0$) has a favorable effect. A branch with $R = 0$ takes over all the current after a certain amount of time, even if other branches have a resistance near zero. This behavior is a good prerequisite for a current-controlled circuit technique. To suppress unwanted reflections additional resistances have to be applied for the termination of superconducting wires. Due to these ohmic resistances the power dissipation of a superconducting wire is not zero.

### 14.3.3 SQUID - Superconducting Quantum Interferometer Device

A *SQUID* (superconducting quantum interferometer device) is a well-known basic circuit consisting of a superconducting ring with two Josephson tunneling devices as shown in Fig. 14.10. Due to the magnetic flux a coupling and therefore an interference exists between both Josephson tunneling devices. This effect shows quantum characteristics in a macroscopic structure.

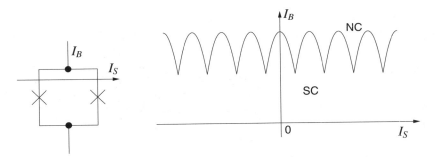

**Fig. 14.10.** SQUID: State diagram of two parallel-connected Josephson elements

Taking the characteristic of Fig. 14.7 one can derive the characteristic of a SQUID as denoted in Fig. 14.10. If the applied field in one element is increased by the control current it is reduced by the same amount in another element. The superposition of both characteristics results in a regular characteristic in the state-chart. This device can be used as a memory cell. Due to the sensitivity of the device another application is the precise measurement of magnetic fields.

## 14.4 Flux Quantum Device

Superconducting memory cells and logic circuits can be realized in such a way that only one magnetic flux quantum is moved per operation. Taking this is into account we get quantum electronics with superconductors. The size of a flux quantum is only $2\,mVps$, hence the switching time of the gates must be in the domain of picoseconds if sufficient voltage signals are to be achieved. Such circuits have been made and tested with frequencies up to $100\,GHz$. The high performance is a crucial advantage for this kind of superconductive circuits.

### 14.4.1 LC-Gate

An interesting approach for superconducting coils is the so-called $LC$ logic. The basic structure is an oscillatory circuit as shown in Fig. 14.11. A switch can short-circuit the coil or split off the capacitor.

Such an arrangement can be realized with superconducting devices. First we consider the case that the information is stored as charge inside the capacitor. If the switch is closed the electrical energy is converted into magnetic energy and back again into electrical energy. With it the voltage changes its sign. If the switch is opened when the current is zero the circuit changes to the other binary state. It seems that no energy is needed during the change of its state. A very similar process occurs if we look at the magnetic flux inside the coil, see Fig. 14.11b.

Is there really no energy loss in this case? Energy is needed for measuring the voltage and the current inside the ring and secondly for the operation of the switch during the passing through zero as described in Chap. 3. The amount of energy for this is at least $kT\,ln2$. Therefore the LC logic is lossy. The relative loss of energy increases with decreasing size of the LC logic. Our considerations are true until we reach the quantum-mechanical limits where the flux quanta and the elementary charge can be seen as packages. Interestingly, the LC logic is nowadays under discussion for low-power CMOS.

### 14.4.2 Magnetic Flux Quantum - Quantum Cellular Automata

A basic circuit that transfers quantum packages are quantum cellular automata (QCA) with magnetic flux quantum SFQ. The configuration of these devices equals the QCA devices with electrons mentioned in Chap. 10. In this case the electrons are replaced as shown in Fig. 14.12 by flux quanta in two fields of the four potential wells.

The flux quanta can be moved by the application of external magnetic fields. Due to energetic conditions adjacent flux quanta move in opposite directions. Circuits of this kind exhibit a relatively slow performance. Josephson tunneling devices are more attractive because only one flux quantum is needed for changing the state of the circuit.

### 14.4.3 Quantum Computer with Single-Flux Devices

Another interesting approach realizes Qubits with Josephson devices in superconducting rings. Such devices have been successfully built by Mooji from

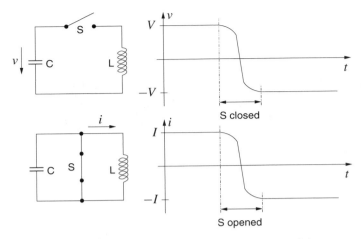

**Fig. 14.11.** LC logic, (a) storage of energy in an electric field, (b) storage of energy in a magnetic field

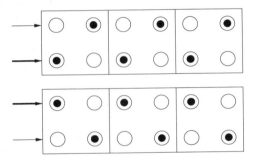

**Fig. 14.12.** QCA with magnetic flux quanta. An applied field can change the position of the magnetic flux quanta

the University of Delft. Here the two states of a Qubit can be expressed with two opposite flux quanta or with two ring currents of opposite directions. The advantages of this technology are the large feature sizes of about $1\,\mu m$ and the possibility to read-out the devices electrically. This simplifies the problem of the I/O interfacing.

If a few such devices are joined together we get a Qubit chain that can perform the operations mentioned in Chap. 5. They can be operated as classical memory cells (Cbits) by slightly modifying their arrangement. In this way an interface to a quantum computer can easily be realized. Figure 14.13 shows the structure of such a computer where the superconducting functional units are controlled by a conventional computer through the Cbit devices. In this case quantum effects occur at the macroscopic level, which is essential for

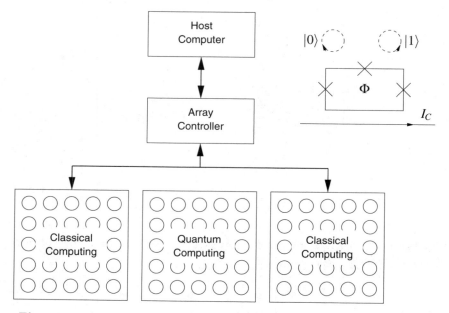

**Fig. 14.13.** Architecture and Qubit element of a superconducting computer

the quantum computer. Therefore the superconducting technology might be a good candidate for such computers.

### 14.4.4 Single Flux Quantum Device - SFQD

A Josephson device stores single flux quanta that can be used for information processing. A possible implementation of such a circuit is depicted in Fig. 14.14. A transmission line with a characteristic wave impedance of $Z \approx 10\,\Omega$ is connected in parallel to a Josephson tunneling device. A bias current forces the circuit to the initial state '0'. An additional impulse raises the current through the Josephson device above its critical current. The tunneling device switches and the operating point moves to '1'. If the impulse vanishes the device returns near to the point '0' without reaching its initial state. Only if the bias current is turned off, will the device return to its initial state. Due to this complex switching process this circuit is comparatively slow. This technique has been investigated in the 1970s.

If the impulse is very short, only a single flux quantum is necessary for the operation. The voltage integral over the time about the resulting voltage impulse is equal to one flux quantum, i.e. the impulse height is $1\,mV$ and the impulse width $2\,ps$. In this borderline case only one flux quantum is used, which is expressed in the name SFQD - single flux quantum device. Taking this idea as a starting point Likarev proposed the RSFQ logic.

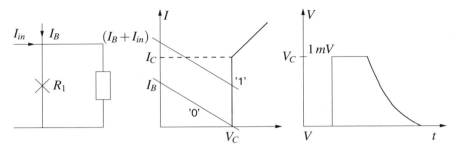

**Fig. 14.14.** Transformation of a few or a single flux quantum into a voltage impulse

### 14.4.5 Rapid Single Flux Quantum Device - RSFQD

A dynamic SFQD is the so-called rapid SQFD as shown in Fig. 14.15. The main difference to the SFQD is the use of an inductance instead of the transmission line. Later a second Josephson device is connected in parallel to the first one. If the first tunneling device changes its state due to an applied impulse the switching process differs from that above-mentioned. The second tunneling device takes over the current delayed in time (effect of the inductance) and prevents the complete switching of the first tunneling device. The circuit is strongly damped to prevent overshooting.

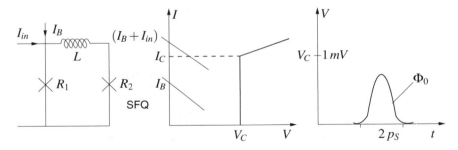

**Fig. 14.15.** Transformation of a single flux quantum into a voltage impulse

Assuming correct sizing of the devices, only one magnetic flux quantum is moved during this operation; it can be sensed at the output node. After completing the operation the circuit returns to its initial state. Due to the avoidance of the normal conduting mode the circuit is very fast.

Figure 14.16 shows a circuit that transfers a direct current signal in an succession of impulses, where each impulse corresponds to a single flux quantum. Applying a direct current the major part of this current flows through the Josephson device $J_3$. If the Josephson device switches, the current distribution inside the circuit changes. The device $J_3$ generates an impulse with

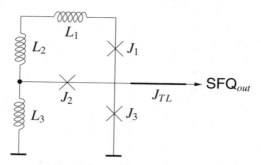

**Fig. 14.16.** Impulse generator

the quantity of one flux quantum. Finally, it returns to the superconducting state and the initial condition is restored. Instead of the inductance and the tunneling devices a transmission line with distributed tunneling devices is used.

Both Josephson tunneling devices are coupled through a superconducting inductance, which minimizes the amount of energy during the switching process. Energy is transmitted in the form of one flux quantum. The flux quantum itself is linked to a pair of electrons, therefore no energy is dissipated.

## 14.5 Application of Superconducting Devices

### 14.5.1 Integrated Electronics

As with transistors logic gates and other digital functional elements can be composed with cryotrons or Josephson devices. The Josephson devices require a new circuit technique with many special features similar to those of circuits fabricated in GaAs. Since the invention of the Josephson device in the 1960s the technology has been improved so that large integrated circuits can be realized. As known from the silicon planar technology this technology is also bases on the deposition of thin layers on a silicon substrate. A commond choice as conducting material is niobium. Isolation layers can be manufactured by evaporating silicon oxide (SiO) or by depositing silicon dioxide ($SiO_2$). The isolation layer inside the Josephson device is made of aluminum oxide with a thickness of only a few nanometers. Corresponding to the planar technology all terminals are located on the surface of the chip.

Many of the large-scale integrated circuits are made for the application of radar engineering and EMC. Their main characteristics are high-frequency operation and low-power consumption together with a high precision. The clock frequency rate is up to $40\,GHz$. Another project was the implementation of a 4 bit microprocessor with 24 000 Josephson devices. With an average feature size of 1.5 $\mu m$ an area of 25 $mm^2$ is occupied. Without any optimization

## 14.5 Application of Superconducting Devices

with regard to the packing density the circuit can be operated at a frequency of up to $1\,GHz$. The necessary memory chips are designed with a capacity of $4\,kbit$. A memory chip of this kind occupies an area of $36\,mm^2$ and exhibits an access time of $0.58\,ns$.

Electrical circuits can also be realized with high-temperature superconductors. For example a shift register made of 32-512 YBaCuO operates at a temperature of $77\,K$ at up to $100\,GHz$. Due to the high working temperatures and the small switching energies special steps have been taken to reduce the failure rate due to thermal fluctuations. At this temperature silicon circuits can also be operated, which allows the integration of CMOS interface circuits at the same time. According to the principle of RSFQ devices different logic gates and more complex gates can be realized. Systems composed of these circuits allow a high-speed operation with clock frequencies of up to $900\,GHz$.

A MOS device operating at low temperatures is interesting for various reasons: The increased mobility of charge carriers due to the reduced collision rate with lattice atoms. Due to decreased diffusion and recombination processes the inverse leakage currents are reduced. The decreased voltage equivalent of thermal energy reduces the Debye length. A temperature of $77\,K$ can easily be achieved with liquid nitrogen. Therefore the application of MOS electronics is beneficial. Despite all these improvements the effort for cooling the circuits is only useful for large data-processing systems.

### 14.5.2 FET Electronics - A Comparison

The most interesting approaches concerning the nanoelectronics are the single-electron transistor and the single flux quantum device. Both devices are working with single quanta hence the borderline of classical electronics is reached.

Thus, the following comparison is of fundamental importance for nanoelectronics. Inside the superconducting electronics the switching energy per switching operation is set by the current and by the inductances. Besides the switches, inductances are the most important devices. The energy is determined by

$$E_{SC} = L\,I^2 = \frac{(n_L\,\Phi_0)^2}{L}. \qquad (14.4)$$

Inside the field effect electronics the main parameters are determined by the voltage and the capacitors. Charging of a capacitance needs the energy of

$$E_{SC} = C\,V^2 = \frac{(n_C\,q)^2}{C}. \qquad (14.5)$$

In (14.4) and (14.5) the energy depends on the number $n_L$ of flux quanta and on the number $n_C$ of electrons. If the error probability caused by thermal noise is equal in both cases the switching energy must exceed the thermal energy $kT$ by orders of magnitude. One has to remember that the operating temperature of both circuits differ considerably. The condition for equal error probabilities can be derived by using the relationship of $n_C$ to $n_L$ according to

$$\frac{n_C}{n_L} = \frac{\Phi_0}{q}\sqrt{\frac{C}{L}\frac{T_H}{T_L}}. \tag{14.6}$$

The term $\sqrt{L/C}$ can be replaced by the wave impedance $Z$. Introducing the free-space impedance $Z_0$ we get

$$\frac{qZ_0}{\Phi_0} = 4\alpha. \tag{14.7}$$

Taking all into account we obtain

$$\frac{n_C}{n_L} = \frac{1}{4\alpha}\frac{Z_0}{Z}\sqrt{\frac{T_H}{T_L}}. \tag{14.8}$$

If we take typical wave impedances between $3-30\,\Omega$ a resistance ratio between $10-100$ is obtained. The ratio of the temperatures is about 100. With it the ratio of the elementary quantities is around $10^5$. On the assumption that both devices are of equal size and with equal error probabilities a switching operation needs $10^5$ times more electrons than flux quanta.

If a device is scaled down by $\alpha$ and assuming $n_L > 1$ the switching energy is reduced by $\alpha^{-3}$. If $n_L = 1$ the switching energy is proportional to $\alpha$. The last-mentioned is valid for the actual dimensions in superconducting electronics so the switching energy is increased when reducing the size of a device. A switching operation with field effect electronics needs more than 100 charge carriers, therefore the switching energy decreases as $\alpha^3$ with the exception of single-electron transistors. In both cases the switching times are reduced by $\alpha^{-1}$. Superconducting electronics consume more energy, which makes it superior to FET logic, which describes the highest possible clock frequencies.

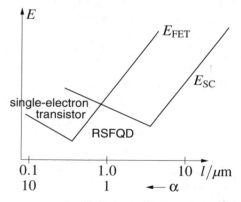

**Fig. 14.17.** Switching energy of FET and SC electronic with regard to the feature size

Superconductivity is not the best candidate for the development of nanoelectronics, because the development of nanoelectronic circuits is mainly set by parameters like the energy dissipation. On the other hand, the standby power dissipation is zero for superconducting electronics, which is of great importance even in large systems.

Interestingly the use of superconducting electronics does not result in a saving of energy. The power dissipation of a circuit is $E_V = m_V k T_L$. The variable $m_V$ is a proportionality factor that includes the complexity of a circuit and the quantity of the switching energy. To cool a superconducting circuit from room temperature $T_H$ to a lower temperature $T_L$ a refrigerating machine must be applied. The machine needs the energy according to the theorem of Carnot

$$E_M = E_V \frac{T_H - T_L}{T_L}. \tag{14.9}$$

The total power dissipation is then set by

$$.E_{all} = E_M + E_V = m_V k T_H. \tag{14.10}$$

The total power dissipation is the exact energy the circuit would consume if operated at the temperature $T_H$ and assuming equal noise immunities concerning thermal fluctuations. With regard to the power dissipation superconducting electronics show a beneficial effect where no extra cooling is needed, e.g. space travel.

The transition from superconducting electronics to roomtemperature electronics is not easy. A good conductor for electrical current is also a good conductor for heat. To get an acceptably high thermal resistance the conductor has to become relatively long or highly resistive. Both conditions deteriorate the signal transfer time and the signal attenuation. A remedy for these problems are optical fibres on the condition of appropriate interface circuits.

Another topic is the reliability of such circuits. Electronics operated at low temperatures is resistant to aging due to reduced diffusion processes. On the other hand, repeated cooling and heating leads to a reduced durability. The difference between the expansion coefficient of the materials causes high mechanical stresses.

### 14.5.3 The Electrical Standards

The investigation of quantum electronic devices is interesting because the 'transistor' parameters are set by fundamental constants. Therefore no parameter scattering exists. Nevertheless, these effects are more important for standard measures than for electronics.

At the moment the following effects can be used for standard measures:

1. A sinusoidal voltage applied to a single-electron transistor causes a current $I$ that is the product of the elementry charge and the frequency. The

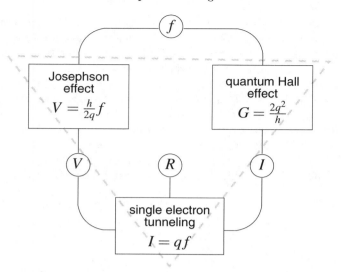

**Fig. 14.18.** The triangle of the quantum measurements for electrical standards: voltage, conductance and current

oscillations of atoms can be used for the determination of physical quantities like time and frequency. Since the frequency can be controlled exactly, such devices can be used as a current standard.
2. A Josephson device can be used as a voltage standard. In this case a voltage $V$ adjusts a frequency, which in turn can be measured exactly.
3. The quantum Hall effect (von Klitzing) combines the quantities voltage and current on the basis of a fundamental constant resulting in a standard for the electrical resistance $R = 1/G$.

Figure 14.18 summarizes the three basic effects with regard to the standard measures. Most of these effects are interesting for metrology but will gain more interest for nanoelectronics. Their main advantage is the lack of parameter scattering.

## 14.6 Summary

Although superconductors exhibit some attractive and interesting characteristics their application in the field of information processing is limited and restricted to some special cases. This is due to the necessity of the very low-temperature ranges that are inherent to superconductivity. Another serious limitation comes from the high action of a flux quantum. Further microminiaturizing will reach the limits of quantum mechanics with the characteristics of high power dissipation and short switching times. Nowadays the direction in the development of superconducting circuits goes towards very high speed information processing in the area of low temperatures, e.g. space electronics.

# 15

# The Limits of Integrated Electronics

## 15.1 A Survey about the Limits

The following describes the most important limits that will restrict the development of micro- and nanoelectronic circuits [45]. The *physical limits* are of a fundamental nature and can not be overcome. This is true, with the reservation that the limits as seen today are not necessarily final ones because the research in physics continues. We can expect considerable progress in the field of quantum computing, which will move some limits far beyond current ones. Some of these physical limits that are presented for classical microelectronics are summarized below.

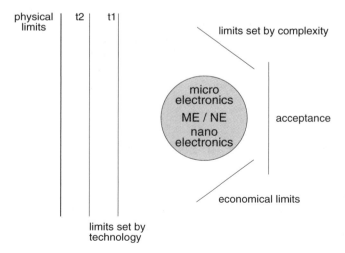

**Fig. 15.1.** Some limits of micro- and nanoelectronics ME/NE. Further development enlarges the circle ME/NE, so the borderline can reach different limits. The technological limit is indicated for the times t1 and t2

As seen in Fig. 15.1 before the circle reaches the physical limits *technological limits* are reached. These limits address topics like device dimensions, chip area, and power-dissipation density [2]. The technological limits are time dependent and can be moved to better values with the development of advanced technologies. On the other hand, the development of new technologies can be too expensive, so economics would restrict further developments.

Thermodynamic effects have a great influence on the characteristics of microelectronic circuits. Due to their thermal energy the particles in a solid are in constant motion. Similar to the spread of an ink drop when it is put into water the concentration gradients inside a solid vanish after a certain period of time. During the compensation diffusion currents appear. Additionally the concentration distribution is subject to fluctuations. All these processes affect the characteristics of a device, i.e. they determine parameters like current flow, noise, heat conduction, doping profiles, parameter spread, and reliability.

In the 1980s one limit was set by the complexity of microelectronic devices. Against all apprehensions this limit could be overcome. New approaches in the field of mathematics and computer sciences have pushed and will probably push this limit further away.

Sometimes the more important limits are set by social constraints. If products are not accepted by the market further development will stop. Products that are assumed to damage our health, e.g. products that emit radio waves, can limit their acceptance on the market. Their acceptance also depends on the costs of the development. If these costs are too high or it is only assumed they could not be paid, the development will be stopped. In all these cases the acceptance of a new technology as a limiting factor is of great importance.

The reflection upon the limits are essential for a company. A product should be optimized permanently if a good market position is required. System performance and reasonable prices are prerequisites for a successful product. The problem during the development of a product is the permanent change of technical and economic conditions. The solution of this optimization problem can be simplified by knowing the limits of the technology. They indicate the available scope of the development of that technology.

## 15.2 The Replacement of Technologies

Business management is of crucial importance in the tremendous improvement of microelectronics. The costs of a transistor or a device have been reduced by orders of magnitude over the past years. While the wafer costs stay mainly constant integration density has changed enormously, see Fig. 15.2. The yield increases due to a reduction of the failure rate and parameter spread. Under these conditions the introduction of a new technology, results in lower costs which in turn drives the further development. From todays perspective the development of silicon technology continuous up to 2010 without reaching any degree of saturation. At this time new technologies with a certain degree of ripeness are launched to market.

The reduction of costs is possibly due to the present human and material resources. Due to the number of competitors, who are subject to the same economic and social conditions, there is an obligation to take part in the development of new technologies if one aims to be a leader in one field of technology. This success is necessary due to the high costs of a chip foundry. At the moment only a few 'nations' are able to invest in such technologies, e.g. Japan, US, and Europe. Because of cost savings the main activity centers are shifting to the Far East.

An alternative is the change of huge chip foundries to a large number of decentralized small ones. The small chip foundries can satisfy the needs of the customers. How far this approach is successful from the technical and economical point of view is still not certain. More probable are small, flexible, and computer-controlled fabs with fast and individual custom design of ICs.

In the future, nanotechnology may create new technological approaches that are cheaper than today's silicon technology. Self-assembling of devices without an expensive lithography may play an important role for the emerging technologies. In this case the replacement of technologies with improved ones can be done sooner.

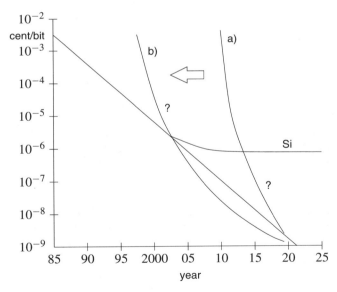

**Fig. 15.2.** Costs per device. The limit of silicon technology is expected for the year 2010. After this, new technologies will emerge (a). If these technologies are cheaper the change of a technology occurs sooner (b). The choice of technology is still uncertain (?)

## 15.3 Energy Supply and Heat Dissipation

For microelectronics, and in future for nanoelectronics, heat generation in integrated circuits is of great importance. Heat conduction can limit the performance of a system. The following idea can be used as a starting point, see Fig. 15.3: A functional block contains $10^{12}$ nanostructured switching elements in a volume of $1\,\mathrm{cm}^3$. If each element requires a switching energy of $1\,\mathrm{fJ}$ and is operated at a clock frequency of $1\,\mathrm{GHz}$ a switching current of $1\,000\,000\,\mathrm{A}$ flows through the block with the dimension of $1\,\mathrm{cm}$. If the devices are operated at a voltage of $1\,\mathrm{V}$ the power dissipation is $1000\,\mathrm{kW}$. Assuming a switching energy of $1\,\mathrm{aJ}$, which seems to be attainable for nanoelectronics, the switching current is still $1000\,\mathrm{A}$. If only $0.1\%$ of the devices are active at the same time the switching current is $1\,\mathrm{A}$. This idea shows the important role of a good system design concerning aspects like heat conduction and low standby power dissipation. Otherwise the obtained values are difficult to manage in technical systems.

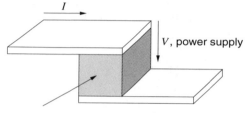

nanoelectronic system, volume $1\,\mathrm{cm}^3$

**Fig. 15.3.** The problem of the power supply of a system with $10^{12}$ devices

Heat conduction is a diffusion process. The power that can be dissipated as heat can be expressed with the diffusion equation:

$$P_{th} = -\kappa \frac{\partial T}{\partial x}. \qquad (15.1)$$

$P_{th}$ is the power of the heat flow, $T$ the temperature, and $\kappa$ denotes the thermal conductivity, which corresponds to the diffusion constant. Assuming a device with the temperature $T_1$, the thermal resistance of $R_{th}$, and an ambient temperature $T_2$ we can write (15.2) according to Ohm's law:

$$P_{th} = \frac{(T_1 - T_2)}{R_{th}}. \qquad (15.2)$$

The ambient temperature and the maximum working temperature are given, so the thermal resistance $R_{th}$ determines the heat flow, i.e. the heat dissipation.

15.3 Energy Supply and Heat Dissipation    249

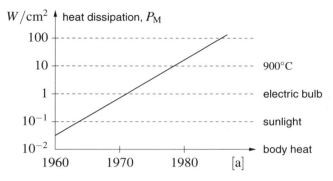

**Fig. 15.4.** Heat dissipation $P_M$ of a module

The power dissipation has its originate in the transistor itself. From there the power dissipation or heat has to pass through the chip and the module before the working environment is reached. Next to the improvement of microelectronics the heat conduction could also be improved, as denoted in Fig. 15.4. At the beginning the improvement was due to improved printed circuit boards and packaging. Heat conduction $P_M$ could be increased to $1\,\text{W}/\text{cm}^2$. The application of a ventilating fan increased this value to $10\,\text{W}/\text{cm}^2$. The next step in the development was fin cooling where the chips on a module are cooled with water from the backside. The backside exhibits a number of grooves that are fabricated by micromechanical structuring. This results in a $P_M$ of about $100\,\text{W}/\text{cm}^2$, which is a typical value for high-performance microcomputers.

During the design of a microelectronic system the context between heat conduction $P_M$ of a module and the technology-dependent power-delay product of a gate is of great importance [45]. This context is explained with the help of the following simple derivation. Let us assume a module containing $n$ integrated devices as shown in Fig. 15.5. The module area $A_M$ containing the chip must be large enough to emit all the heat dissipated by the active device $n_a$ on the chip and can be expressed by

$$A_M \overline{P}_M > P_{Va} n_a. \tag{15.3}$$

This assumption limits the maximum temperature of the chip. The signal delay $t_m$ between two chips should be smaller then the delay time $t_d$ of a gate. This leads to

$$t_d > m_M \frac{1}{\sqrt{A_M} v_M}. \tag{15.4}$$

The term $m_M$ is empirical and $v_M$ denotes the propagation speed of the signals inside the module. If $A_M$ is eliminated from (15.3) and (15.4) we get the following:

$$t_d^2 > \frac{m_M^2 n_a}{v_M^2 \overline{P}_M} P_{Va}. \tag{15.5}$$

## 15 The Limits of Integrated Electronics

This equation determines the limit inside the power-delay diagram of Fig. 15.6 due to heating. This diagram contains the characteristics of different integration levels and for a heat conduction of $100\,\text{W}/\text{cm}^2$. The higher the delay time the higher can be the switching energy.

Another limit follows from the heat dissipation of a driver circuit to the chip, see Fig. 15.5. The power dissipation $P_{VL}$ of a power gate may not exceed the maximum allowable heat dissipation of a chip. The heat inside the chip should be uniformly distributed,

$$A_C \overline{P}_C > P_{VL}. \tag{15.6}$$

Once again $m_c$ is determined empirically, $C$ is the distributed capacitance and $\sqrt{A_C}$ the assumed wire length. If the chip area is eliminated we get

$$P_{VL}\, t_{DL} = m_C \frac{1}{2} \sqrt{A_C \overline{C}}\, V_B^2. \tag{15.7}$$

$$P_{VL}\, t_{DL}^2 < \frac{1}{4} \frac{m_C^2\, \overline{C}^2\, V_B^4}{\overline{P}_C}. \tag{15.8}$$

If we equate the switching time $t_{DL}$ of the driver with the gate delay $t_D$ the limiting condition of (15.8) can be drawn on a $P_V - t_D$ chart.

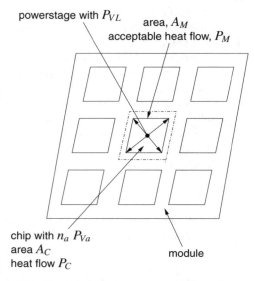

**Fig. 15.5.** Heat dissipation on a module and on a chip

A fundamental restriction of aggressive scaling could be the heat conduction especially in those cases where the power dissipation per unit area increases. Today this problem is not of great interest in information-processing circuits. Nevertheless it is an important topic for power stages.

If these reflections are passed onto integrated circuits and taking the switching time into account we get the chart of Fig. 15.6. This chart contains the limits for field effect transistors. The removal of heat is actually restricted to values around $100\,\mathrm{W/cm^2}$. The more devices integrated on a chip the lower must be the switching energy. Lowering the switching energy causes smaller dimensions, therefore the performance of the devices increases. On the other hand, the absolute limit of lowering the switching energy is set by thermodynamics of about $50\,kT$. The uncertainty relation leads to the quantum-mechanical limit, which is important for single-electron transistors.

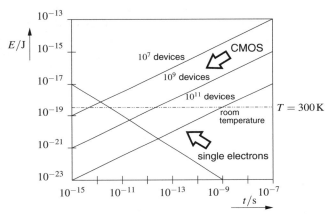

**Fig. 15.6.** Switching energy versus delay for SET and CMOS devices

Also important are the consequences of further scaling down the energy supply of the circuits. Even if scaling reduces the supply voltage to $V/\alpha$, the currents on the supply lines increase due to an increase in the number of switching devices. From this the well-known problems for power supplying integrated circuits with low voltages and high currents occur. Three typical effects arise: Due to high currents a voltage loss on the supply lines occurs. The losses have a great influence when operating the circuits at low voltages. A remedy for this problem are distributed power-supply lines. The high current densities can cause a migration of material which in turn leads to circuit failures. Due to unavoidable line inductances each current variation causes a voltage impulse on the supply lines, which in turn can reduce the reliability of the system. Remedies for this problem are supply lines with a lower inductance per unit length, blocking capacitors, and circuit techniques that avoid high current pulses.

## 15.4 Parameter Spread as Limiting Effect

Due to the diffusion of particles, layers with different electrical characteristics can be built. On the other hand, diffusion is not a equally distributed process, which can cause problems inside an electrical circuit. A very important application of the diffusion process is the production of integrated circuits. As an example the production of a pn-junction in a planar technique is explained. The starting point is a p-type substrate that is covered with a $SiO_2$ layer. The part of the $SiO_2$ layer that is opened by an oxide window is covered with n-type phosphorus atoms, see Fig. 15.7a. If this structure is exposed to temperatures above 1000°C the phosphorus atoms diffuse into the p-type substrate, which results in a doped layer with a different polarity. In addition, Fig. 15.7 shows that some of the particles diffuse under the adjacent oxide (underdiffusion). When the diffusion process is maintained the layer continues growing. The doping profile depends on the time and is shown in Fig. 15.7b and can mathematically be formulated by (15.9).

$$n = \frac{Q}{\sqrt{\pi D t}} e^{-x^2/(4 D t)}. \tag{15.9}$$

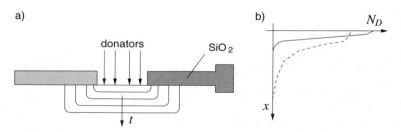

**Fig. 15.7.** Process of diffusion for the realization of semiconductor structures. (a) Diffusion process in a semiconductor and (b) the diffusion profiles as a function of time

In (15.9) $Q$ denotes the initial density of phosphorus atoms and $D$ the diffusion constant. If the temperature is lowered the diffusion process does not stop but reduces its speed. This also indicates that the electrical parameters that are adjusted during the manufacturing process can vary with time and can sometimes cause device failures. Therefore the diffusion process is very important when the reliability of a device or system is of concern.

The equal-probability distribution of (15.9) is only a mean value. The other values of each pn-junction vary around this value. This results in a spread of the electrical and technological characteristics of the devices. This is a fundamental effect for microelectronics and is mostly detrimental to integrated circuits.

## 15.4 Parameter Spread as Limiting Effect

In the following we estimate the parameter spread of the diffusion process with the help of a simple example. The starting point is a volume $V$ that is divided into $N$ cells each containing $n$ particles. We now ask about the probability $P$ that $n'$ particles stay in one half of the volume. From experience we know that on the average $n'$ equals $n/2$. Due to thermal motion sometimes more and sometime less than $n/2$ particles are inside half of the volume. To derive the spread mathematically we calculate the number $m$ of possibilities to put $n'$ particles in one half of the volume. Eqation (15.10) can be derived with methods of combinatorial analysis:

$$m(n') = \frac{(\frac{N}{2})!}{n'!(\frac{N}{2}n')!} \left(\frac{n}{N}\right)^{n!} \left[1\left(\frac{n}{N}\right)\left(\frac{N}{2}n!\right)\right]. \tag{15.10}$$

In the case of $n = N/2$ (15.10) can be simplified to:

$$m(n') = \frac{(\frac{N}{2})!}{n'!} 0.5^{N/2}. \tag{15.11}$$

This equation is evaluated for a very small number of particles $N$ in Fig. 15.8. The higher the number of cells and particles the higher the number of possibilities. The maximum number of possibilities occurs if half of the particles are inside one half of the volume. This state occurs with the highest probability, which agrees with our experience of technical systems.

If the characteristic curves are standardized on the maximum value of $n/2$ we get the chart of Fig. 15.8b. It follows from this that the normalized standard deviation decreases with an increasing number of particles,

$$\frac{\sigma}{\mu} = \frac{\sqrt{0.5 N \frac{n}{N}\left(1 - \frac{n}{N}\right)}}{0.5 N \frac{n}{N}} \approx \frac{1}{\sqrt{n}}. \tag{15.12}$$

This result explains why the spread of the values can be neglected in the field of classical thermodynamics. According to the Avogadro constant $L$ one mole consists of $6.022 \times 10^{23}$ molecules. At the temperature of $0°C$ and pressure of $10^5$ $Nm^{-2}$ the volume is $22\,414\,cm^3$. In this case the scattering region is about $10^{-10}$.

The relations in microelectronics are quite different. If (15.12) is applied to the density of a doped region in a device and if we calculate the spread due to the diffusion process we get the numerical examples of Table 15.1.

Table 15.1. Numerical examples

| Relevant volume | $1\,cm^3$ | $1\,\mu m^3$ | $0.1\,\mu m^3$ |
|---|---|---|---|
| Mean number of doping atoms | $10^{16}$ | $10^4$ | $10$ |
| Normalized standard deviation | $10^{-8}$ | $10^{-2} = 1\%$ | $10^{-0.5} = 30\%$ |

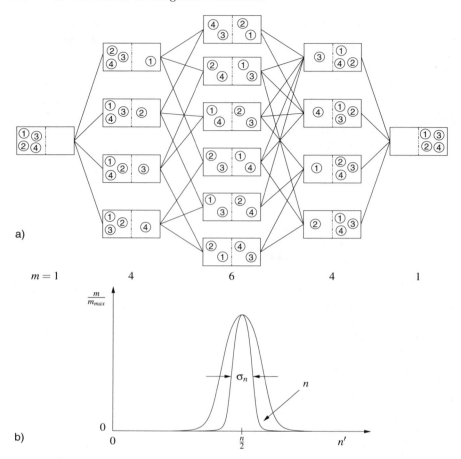

**Fig. 15.8.** Number $m$ of possibilities to place four particles according to (15.11) (a) and for $n$ particles for $n'$ particles per volume (b)

This example demonstrates that with decreasing volume and therefore with a decreasing number of particles the normalized scattering increases. This spread influences various parameters of a device. The threshold voltage of MOS transistors is mainly set by the doping concentration of the space-charge region under the gate. However, the number of doping atoms decreases in the space-charge region only with a power of 2 and not with 3 according to the scaling factor. Nevertheless, this effect seems to become more and more important when scaling down the devices, maybe this is an insuperable limit of classical microelectronics.

Equation (15.13) denotes the depth of the space-charge region depending on the surface potential:

## 15.4 Parameter Spread as Limiting Effect

$$x = \sqrt{\frac{2\varepsilon\varphi_S}{N_A q}}. \tag{15.13}$$

The threshold voltage of a MOS transistor is

$$V_T = \frac{\sqrt{q\,2\,\varepsilon\,\varphi\,N_A}}{C_{ox}}. \tag{15.14}$$

The variation of the threshold voltage depends on the change of the doping concentration:

$$\frac{dV_T}{V_T} = \frac{dN_A}{N_A}. \tag{15.15}$$

According to (15.12) the variation region can be written as follows

$$\frac{\Delta N_A}{N_A} = \frac{1}{\sqrt{N_A x^3}}. \tag{15.16}$$

Equation (15.16) is inversely proportional to the root of the number of doping atoms. With this relation we can reformulate

$$\frac{\Delta V_T}{V_T} = \frac{1}{\sqrt{N_A x^3}}. \tag{15.17}$$

The variation $\Delta V_T$ of (15.17) increases with a decreasing channel region and with lower doping concentrations.

According to *Mead and Conway* [1] the failure probability of an inverter operated at the supply voltage of $V_B$ can be stated to be

$$P_I = e^{-2V_B/\Delta V_T}. \tag{15.18}$$

The probability decreases with higher supply voltages and with a smaller scattering. The larger the transistor dimensions the smaller the spread of parameters. If the voltage can not be decreased the power dissipation can not be diminished as one would like.

We illustrate this effect with the help of an inverter chain. The threshold voltage was lowered to $V_{T0}$ by implantation. Due to fluctuations of the particle density the threshold voltage varies for each transistor, see Fig. 15.9. If the supply voltage of the inverter chain is $V_B$ all inverters with a threshold voltage $V_T$ above $V_B$ do not switch. The probability $P_1$ for this condition follows from the shaded area of Fig. 15.9. The following numerical example is derived from Keyes: A MOS transistor with the dimension of $(0.1 \times 0.5)\,\mu\mathrm{m}^2$ contains about $1000 \pm 35$ doping atoms inside the space-charge region. This variation of the doping concentration results in a variation of the threshold voltage of $\pm 17\,\mathrm{mV}$. If a scattering of $\pm 51\,\mathrm{mV}$ is permitted, 40 of the 1 million transistors operated at 2 V would fail and with it the circuit.

From the probability of a failure the yield for the inverter chain can be stated to be

$$Y_I = 1 - P_I. \tag{15.19}$$

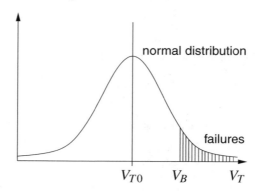

**Fig. 15.9.** Impact of the parameter spread on the yield and reliability: If the threshold voltage is above the power supply voltage the inverter does not work. The shaded area denotes the probability of a failure

For an inverter chain with $m$ stages the yield is considerably smaller,

$$Y_{St} = Y_I^m = (1 - P_I)^m. \tag{15.20}$$

If one wants to achieve a given yield, $P_I$ has to become sufficiently small. According to Fig. 15.9 the supply voltage $V_B$ has to be increased. The minimum supply voltage is therefore not set by the nonlinear behavior of the characteristic curves but by parameter variations. Mead calculated this voltage for a given FET technology and for $10^7$ inverters to be 700 mV. The problem of a reduction of the yield becomes more and more important for devices with decreasing feature sizes.

Although some experts see this problem as insuperable, engineers have found some answers to this problem, as stated in Chap. 2. The solution to this problem can be achieved by two approaches: The first approach wants to reduce the variation region by applying local active process steps. The second approach uses appropriate circuit techniques to avoid the influence of parameter variations to the switching performance. Alternatives are the following two approaches [67]:

- The channel is implemented into an undoped layer that is grown epitaxially upon a highly doped layer. Inside the undoped layer no scattering occurs and the scattering in the doped layer has almost no effect on the spread of threshold voltage of the transistor.
- A futuristic approach is a modified raster scanning microscope that can implant single atoms. However, the time for processing a whole wafer with a diameter of 30 cm is unrealistically long.

## 15.5 The Limits due to Thermal Particle Motion

These stochastic processes result in temporal and spatial fluctuations. Parameter variations are related to spatial fluctuations. Temporal fluctuations arise due to noise, the extension of a doped region, or due to the reliability. All these effects have a limiting effect on microelectronics.

### 15.5.1 The Debye Length

The diffusion of charge carriers can cause a potential difference inside a semiconductor. As an example, an inhomogeneous doped semiconductor is investigated. Due to the inhomogeneous doping profile diffusion of mobile charge carriers occurs resulting in an electrical field within the semiconductor. At the thermodynamic equilibrium the currents due to diffusion and due to the electric field must be equal at each point within the semiconductor. This condition can be explained with the differential equation

$$q D \frac{\partial p}{\partial x} = p q \mu_p E. \tag{15.21}$$

The potential $\phi$ of each point inside the semiconductor can be found by integrating (15.21). With it the voltage between two points can be calculated:

$$V_{12} = (\varphi_1 - \varphi_2) = V_T \ln\frac{p_2}{p_1}. \tag{15.22}$$

This voltage is independent of the doping profile $N_A(x)$ within the semiconductor. The potential along the semiconductor can be derived from the solution of the Poisson equation:

$$\frac{\partial^2 \varphi}{\partial x^2} = \frac{1}{\varepsilon_0 \, \varepsilon_{si}} q \, (p - N_A) \tag{15.23}$$

In the case of an intrinsic semiconductor and with $p = n_i \, exp(\phi/V_T)$ we get (15.24) after standardization:

$$\frac{\partial^2 \left(\frac{\varphi}{V_T}\right)}{\partial x^2} = \frac{1}{L_D^2}\left[e^{\varphi/V_T} - \frac{N_A}{n_i}\right]. \tag{15.24}$$

The term $L_D$ is the so-called *Debye length*. It determines the standardization of the x axis and can be stated to be

$$L_D = \sqrt{\frac{\varepsilon_0 \, \varepsilon_{si} V_T}{q \, n_i}}. \tag{15.25}$$

The lower the temperature and the smaller the relative permittivity $\varepsilon_{si}$ the smaller the Debye length $L_D$. The Debye length is a first indicator for the minimal extension of solid-state devices. Also it determines the constant in the exponential term and describes the fading away of disturbances within the semiconductor. At room temperature and for silicon the Debye length is around 100 nm. An example of a disturbance is the spread of a space-charge region. According to Schottky a space-charge region ends abruptly.

### 15.5.2 Thermal Noise

The thermal motion results in an increased resistance and in a noise voltage that can be measured along a semiconductor. The noise voltage is usually derived from the harmonic oscillator, which is excited by the thermal energy. Such an oscillator is a model for a short-circuited line with the wave impedance $Z = R$. Here a simplified, quantitative derivation is shown. We calculate the thermal noise power at an ohmic resistor, which can be measured with a bandpass filter as shown in Fig. 15.10.

According to the *Nyquist* theorem the signal must be sampled with $\Delta t < 1/(2\,\Delta f)$ to obtain a complete waveform. $\Delta f$ corresponds to the bandwidth of the bandpass filter. In the case of a noisy resistor we observe the power and not the voltage. From thermodynamics it follows that for each degree of freedom, and therefore for each amount of energy, the average energy of $\Delta E = 1/2\,kT$ is related. From the viewpoint of information theory the mean transmissible power is

$$\overline{P} = \frac{1}{2}kT\,2\,\Delta f = kT\,\Delta f. \tag{15.26}$$

The mean noise voltage at a noise-free resistor $R$ can be calculated with the help of the four-pole theory according to

$$\overline{v}^2 = 4\,R\,k\,T\,\Delta f. \tag{15.27}$$

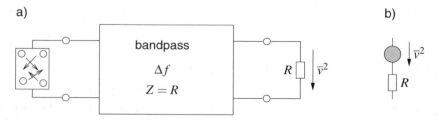

**Fig. 15.10.** Noise voltage of an ohmic resistor and corresponding equivalent electronic circuit: Noise source and noiseless resistance

The noise voltage is proportional to the resistance, the temperature, and the bandwidth. In general, this voltage has a negative effect for electronic circuits and particularly in the processing of analogous signals. A figure of merit for the characterization of the noise voltage is the signal-to-noise ratio. In addition the noise voltage determines the lowest voltage inside an electronic circuit. If a circuit is scaled down the switching energy and therefore the signal power becomes lower. The effect of the thermal power increases with the bandwidth.

Also the minimum switching energy due to thermal fluctuations has to be taken into account. This energy was estimated to be $50\,kT$, see Chap. 9 and Chap. 3.

## 15.6 Reliability as Limiting Factor

The distribution of the parameters as shown in Fig. 15.9 is only valid for the initial state of the devices. Devices with a threshold voltage $V_T$ above $V_B$ fail. As a result of this the device yield is reduced. During the operation of the devices the distribution widens, which in turn causes additional failures. These failures determine the reliability of the devices. The reliability describes the average time a device is able to work. Once again the diffusion process provides an adequate model for the explanation [68].

The diffusion process always occurs at temperatures above $0\,\mathrm{K}$ and increases with rising temperatures. The latter is advantageously used in the manufacture of integrated circuits. If a device is operated at working temperature the diffusion process is small enough that the lifetime of a device is hardly reduced. On the other hand, some failure mechanisms are based on the diffusion process. The lapse of time for this mechanism is very fast so further unwanted failures occur.

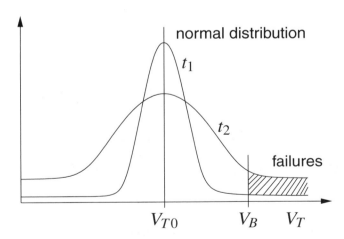

**Fig. 15.11.** With increasing time $t_1 \to t_2$ the failure rate increases. This is due to the enlargement of the distribution caused by the diffusion

Figure 15.11 is based on a normal distribution of the parameters that can be described by (15.28). Instead of $x$ we take the normalized threshold voltage. The distribution widens with time, so the number of failures increases. With (15.9) the time constant for this process can be approximated to

$$\tau_z = \frac{x_0^2}{D} = C' e^{E_A/(kT)}. \tag{15.28}$$

In (15.28) $x_0$ is a fixed quantity. In the above-mentioned example $x_0$ corresponds to $V_{B0}$. If the diffusion constant $D$ is replaced with the result of (3.37) we get the right term of (15.28). Once again $C'$ is a constant and $E_A$ the activation energy.

Now we compare this time constant to the time constant that describes the failure rate in relation to time. In the reliability theory a first approximation of the failure rate is described by an exponential function:

$$N = N_0 \, e^{-\lambda t}. \tag{15.29}$$

$N_0$ describes the number of devices at the initial state, $N$ the number of devices during operation, $t$ the time, and $\lambda$ the failure rate. This equation can be derived from the differential equation with the condition that the relative failure rate per unit time is constant:

$$\frac{1}{N}\frac{dN}{dt} = -\lambda. \tag{15.30}$$

We now consider the term $1/\lambda$ as a time constant, which can be proportionally set to $t_Z$. This can be done because the processes that cause failures are of equal nature to those that are responsible for the widening of the parameter distribution with time. Equation (15.31) can be derived if $\lambda$ is set proportional to $1/t_Z$,

$$\lambda = C_A e^{-E_A/(kT)}. \tag{15.31}$$

If the diffusion process is taken as a basis we can regard the failure rate $\lambda$ as a function of the diffusion constant $D$ in which the activation energy $E_A$ is the dominant quantity. Therefore the activation energy $E_A$ is a characteristic measure in microelectronics for describing failure mechanisms. $C_A$ includes different quantities.

If the failure rates $\lambda_1$ and $\lambda_2$ are measured at two different temperatures $T_1$ and $T_2$ we get the activation energy $E_A$ from (15.31). A graphical interpretation of this is shown in the so-called *Arrhenius Plot* of Fig. 15.12.

$$E_A = kT_1 \frac{T_2}{T_2 - T_1} ln \frac{\lambda_2}{\lambda_1}. \tag{15.32}$$

Most of the device-failure mechanisms have an activation energy around 1 eV. In practice usually more than one failure mechanism occurs at the same time. These failure mechanisms have different activation energies. A superposition of different exponential functions can be measured.

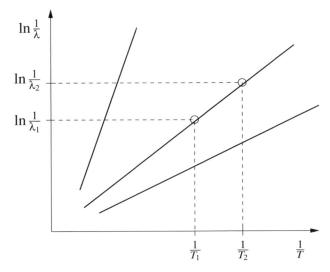

**Fig. 15.12.** Dependence of $\lambda$ of the temperature $T$ and the activation energy $E_A$. $E_A$ can be determined by two measurements with different temperatures

If (15.30) is converted into a difference equation

$$\Delta N = -\lambda N \, \Delta t \tag{15.33}$$

and if $\Delta N = 1$ we get the time $\Delta t$ that passes until a failure on the average will occur,

$$\Delta t = \frac{1}{N \lambda}. \tag{15.34}$$

A device with the failure rate $\lambda$ stated in hours will fail on average in $1/\lambda$ h. The time until a failure in a system with $N$ devices occurs is reduced with increasing $N$.

The following simple example shows that a large number of devices complicates the guarantee of a long life time and the reliability. The *mean time between failure* (MTBF) which is set to $1/\lambda$ is around 1 000 a for todays complex microelectronic circuits. In case of a digital clock with only one chip this number results in an outstandingly good reliability. On the other hand a mainframe memory with 10 000 such components will fail on average each 0.1 a due to a faulty component. For a computer system this is an unacceptable number. A remedy for this is an improved level of quality.

Now we assume that each memory chip contains $10^7$ devices. If we assume an MTBF = 1000 a the MTBF of a single device is $1/\lambda = 10^{10}$ a, see Fig. 15.13. This time corresponds more or less to the age of the universe. Even on this reliability level it is not simple for the manufacturer to determine the reliability of his devices. Without any insight into the failure mechanisms of the devices the reliability can not be improved any further. In addition, the experience of tested devices in real-world application is of crucial importance for increasing the quality.

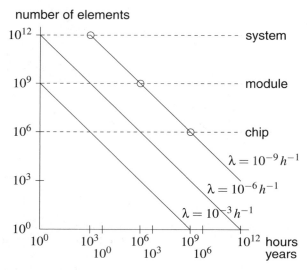

**Fig. 15.13.** MTBF in hours or years and the number of devices on the module or chip level for different failure rates $\lambda$. The markers correspond to the examples in the text

A product has to function for many years. However, a lifetime above 10 years is not worth aspiring to due to increased costs. On the other hand, failures that occur in the first year of operation are troublesome. The reliability decreases from the system level down to the device level. To find the correct values at each level is not an easy task.

Redundancy is a good way to increase the reliability on the hardware level. In contrast to software, where the concept of redundancy can be easily fulfilled, additional control logic for switching-on the redundant parts is area consuming and the logic is also subject to failures. Therefore fault-tolerant architectures at higher integration levels are of great importance, especially for microelectronics of the next generation.

In the regime of nanoelectronics the discussed problems might be something different: The characteristics of quantum electronic devices are determined by physical constants, which in turn do not scatter. In the inside of molecules no diffusion processes occur, so no reliability problem exists. On the other hand, the issue of reliability occurs when regarding compounds, necessary contacts, and supporting material. These problems are partly unsolved.

The same context as for reliability is true for the time: The results of any computation should be computed in a few hundredths or tenths of a second.

A faster computation is of no interest for human beings but longer calculation times have an effect on the impatience and reduce the effectiveness of his work.

In practice, the limits of thermodynamics are of great priority but for fundamental considerations physical limits are more expressive.

## 15.7 Physical Limits

The preceding effects of classical physics permit a derivation of physical limits of micro- and nanoelectronics. In the following discussion the most important limits from thermodynamics, the theory of relativity, and from quantum mechanics are presented. A graphical overview of these limits is shown in Fig. 15.14. For a better orientation the straight-line characteristic for $1\,\text{fJ} = 10^{-15}\,\text{W s}$ is shown in the power-delay chart. This number is characteristic for current integrated circuits. The classical laws of physics that have been under discussion for a long time are not necessarily definite limits. Without any doubt they are important in the periphery of nanoelectronics.

Some of these limits might be overcome in the future, the more as quantum-mechanical aspects are used. When regarding the quantum computer the results of any computation are valid without any delay in time when applying the appropriate input signals and after a unitary transformation. Some people believe that a signal propagation as fast as the speed of light is possible in tunneling devices. Other theories suggest that information processing can be done with a negligible amount of energy.

All these ideas should be taken with caution because new concepts for computers such as those for the quantum computer could affect the limits presented below.

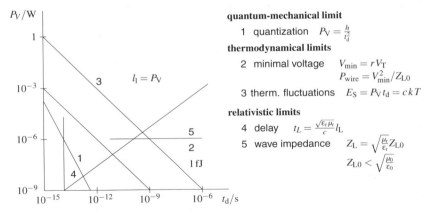

**quantum-mechanical limit**
1 quantization  $P_V = \frac{h}{t_d^2}$

**thermodynamical limits**
2 minimal voltage  $V_{\min} = rV_T$
   $P_{\text{wire}} = V_{\min}^2/Z_{L0}$
3 therm. fluctuations  $E_S = P_V t_d = ckT$

**relativistic limits**
4 delay  $t_L = \frac{\sqrt{\varepsilon_r \mu_r}}{c} l_L$
5 wave impedance  $Z_L = \sqrt{\frac{\mu_r}{\varepsilon_r}} Z_{L0}$
   $Z_{L0} < \sqrt{\frac{\mu_0}{\varepsilon_0}}$

**Fig. 15.14.** Physical limits for microelectronics with regard to power and delay

## 15.7.1 Thermodynamic Limits

Thermal movement of charge carriers impedes the current flow in a switching device. The dependence of the current through a pn-junction due to an applied voltage is reduced at higher temperatures. The higher the temperature the smaller the curvature of the diode characteristic. If we want to take advantage of the nonlinear behavior of the device the working voltage should not be too small. The smallest value of this voltage is

$$V_{min} = r\,V_T = r\,k\,T/q, \qquad (15.35)$$

with $r = 4 - 8$. The higher the temperature $T$ the higher the voltage $V_{min}$.

To reduce scattering inside the device during a switching event dozens of electrons should be involved. Therefore the lower limit for this value $W_S$ is about $25\,k\,T$. Actually this value is mostly too small, so thermal fluctuations would occur inside the device.

In Sect. 9.1 the effect of thermal fluctuations on switching devices was discussed. We derived the probability $P_{th}$ that energy values due to thermal fluctuations are above the switching energy $W_S$. In the following we define the failure rate that is small enough for real applications. From this failure rate the lowest possible switching energy can be derived if the operation time $t_0$ and the switching time $t_S$ are taken into account. From the number of possible switching processes $n = t_0/t_S$ and with $P_{th}$ we get the failure rate:

$$R_{th} = \frac{t_0}{t_s} P_{th} = \frac{t_0}{t_S} e^{-W_S/(k\,T)}. \qquad (15.36)$$

Now we assume $10^6$ switching devices with a switching time $t_S$ of 1 ns. If only 1 failure per year (1 a = $31.5 \times 10^6$ s) is allowed $W_S$ has to be above $50\,k\,T$. This value can be stated for the thermodynamic limit shown in Fig. 15.14 [69].

Another thermodynamic limit is a result of diffusion. On the one hand this effect is applied in the manufacturing of integrated circuits, on the other hand, diffusion occurs during the application of the devices, which causes a change in some electrical parameters. These changes can be large enough that the device will fail over its lifetime. In addition, the supply of energy and heat removal sets further limits.

## 15.7.2 Relativistic Limits

Some important limiting effects for microelectronics can be derived from the theory of relativity. If information is coupled to discrete energy quantities we must assume that these quantities can not propagate faster than the speed of light $c$. The switching times can not be reduced below a certain limit because the devices can not be scaled down as one would like. Let us assume that the active area of classical devices can not be smaller than the Debye length, then we get, with (15.25), a signal transfer time $t_L = 1$ fs. This value is also shown in

Fig. 15.14. From this idea we can also derive that the core of a microprocessor for future supercomputers can maximally have the volume of a sphere with the radius of 2.5 cm. This size follows from the fact that a light wave can cover the distance of 3 cm in 0.1 ns, which corresponds to a frequency of 10 GHz.

Another limiting effect for microelectronic devices is the wave impedance of a transmission line, which can not be larger than $Z_0 = 377\,\Omega$. This finite wave impedance and the lowest possible voltage level $V_{min} = r\,V_T$ (derived from thermodynamics) cause a power dissipation in the terminating impedances which can not be further reduced. With (2.11) we can calculate the power dissipation as

$$P_{VL} = \frac{V^2}{Z_0} = \frac{\varepsilon_0\,c\,k}{q^2}T^2 r^2. \qquad (15.37)$$

Equation (15.37) depends only on the temperature $T$, which is also depicted in Fig. 15.14. In microelectronics the values for the signal delay time, the wave impedance, and the power dissipation belong to the relativistic limits.

### 15.7.3 Quantum-Mechanical Limits

The interaction of light with a solid can only be explained by quantum mechanics. Regarding (3.6) in the case of $n = 1$ we get the energy per switching event according to

$$W_S \geq \frac{h}{t_s}. \qquad (15.38)$$

Or for the power:

$$P_V \geq \frac{h}{t_s^2}. \qquad (15.39)$$

These values can not be reduced any further. They are termed the quantum-mechanical limit in Fig. 15.14.

In the same way, failures of switches occur due to thermal fluctuations, failures due to tunneling of charge carriers inside the switches is possible. Tunneling is an effect of quantum mechanics. In the following a rough estimation of this type of failure is given.

### 15.7.4 Equal Failure Rates by Tunneling and Thermal Noise

The derivation of the failure probability due to thermal fluctuations assumed that a failure occurs each time the energy of a charge carrier is above $W_S$. If we consider failures by tunneling, the probability that a charge carrier can tunnel through a barrier is dependent on the energy $W_S$ of the barrier and on the depth $d$. A first approximation for this probability is shown in (12.1) with $P_{tu} = D$. $\overline{E}$ is the mean kinetic energy of the electrons.

Now we estimate the probability that failures due to tunneling and thermal fluctuations equal. With the assumption $P_{th} = P_{tu}$ and with (3.30) and (9.2) we get:

$$\frac{W_S}{kT} = \frac{2d}{\hbar}\sqrt{2m\left(W_S - \overline{E}\right)}. \tag{15.40}$$

If the mean kinetic energy $\overline{E}$ of the electron is half the energy $W_S$ of the barrier we get the first-order estimation:

$$d = \frac{\hbar\sqrt{W_S/m}}{2kT}. \tag{15.41}$$

If we assume $W_S = 50\,kT$ the depth of the barrier can be calculated as $d = 2\,\text{nm}$ at $T = 300\,\text{K}$. At this depth the failures caused by tunneling equal those caused by thermal fluctuations. Therefore the dimension of a device is a limit for the microminiaturization of classical devices. This limit is almost reached by semiconductor memories where the oxide thickness is around 5 nm. The charge flow due to tunneling has to be taken into account.

## 15.8 Summary

The drift current because of an electric field and the diffusion current because of density differences are fundamental effects when the motion of particles is investigated. They are not only responsible for the switching process but also for the reliability of the devices. Due to thermal motion the particle concentration is not constant, which causes variations of the electrical characteristics of the devices. When considering nanoelectronics the problems are of a different nature. The universal constants do not scatter and the molecules do not move. But the wires and the supporting material are subject to the process of diffusion. Reliability is a very import characteristic especially for integrated systems in the regime of nanoelectronics.

The diffusion process also determines the heat conduction, which is elementary for electronics. With the help of simple models some practical limits have been derived.

In addition the physical limits are important for principle considerations. They define the scope for future development of classical electronics. At the moment, integrated electronics are far from these limits.

# 16
# Final Objectives of Integrated Electronic Systems

A book about nanoelectronics should present visions of integrating the information processing machines into nanoelectronics. The main objectives of the effort in the development of integrated electronics are systems for information processing without doubt. From our present view such machines must be as small as possible to gain new information, or to remove uncertainty in our knowledge [70]. These tasks should be fulfilled with as low a power as possible.

## 16.1 Removal of Uncertainties by Nanomachines

The main target of information technology is the removal of uncertainties in our knowledge, i. e. to be well informed. Uncertainty is correlated to information because information is described in terms of the probability of an event. The uncertainty and therefore the probability are caused by a spatial event or are due to complexity. In the first case we can not directly observe an event, whereas in the second case the selection on a specific event is impossible due to the diversity of possible events. One task of information technology is to provide people with appropriate machines for removing these uncertainties. We should not forget that the semantic information is the relevant one for us whereas the technical system only handles numerical information, semantic information does not figure at all.

On the one hand people have a strong demand for exchanging information with other people at any place, and any time. The satisfaction of these demands leads to world-wide networks and portable machines, e.g. laptops and cellular phones with wireless connections to the communication networks. For all these systems electronics have to be lightweight, small, and power saving. This development leads to the terms of *ubiquitous computing* and *ubiquitous networking*.

In addition, there are demands of the human society to process information. First, mankind has to solve important complex problems such as the rate of economic development in future, to consider environmental pollution or to forecast the weather more exactly than today, to organize the health services more efficiently, to note only some examples. All these problems are characterized by though mathematical tasks that are inherent to them. Therefore they can only be solved by high computational power, which present computers do not provide.

Whereas up to now the development of microelectronics has pushed the information technology to better and better performances, the further development of microelectronics will be determined by the needs of the market. The former case is called forward integration, which means that microelectronics intruded into systems or markets, were controlled by technology, i.e. the integration of microelectronics in systems. From now we get a backward integration that yields a wide variety of applications such as voice-controlled computers, image-processing systems, intelligent data mining, automatic programming tools, design boxes, and regulating systems for complex processes, etc.

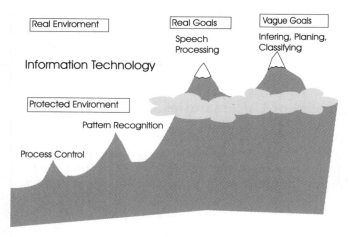

**Fig. 16.1.** Goals of the information technology: Today we do not have an exact view of the complexity so the significance of forward-looking microelectronics is not obvious, H. Martini

Today our forecasts concerning the nanoelectronics are relatively precise, however, the development is just at its beginning (Fig. 16.1). The future aims at a kind of information processing where the information comes from the real world environment containing interferences and not from artificial, and protected systems. Neural nets of inferior animals show a behavior such as this. Therefore biologically inspired learning will influence new techniques for improved information processing. To take living things as an example seems to be reasonable since the techniques from nature may provide us with machines, e.g. robots, with similar features.

One alternative leads to nanomachines based on the concept of the Turing machine. An interesting but more or less academic example is the DNA computer with nanomachines. Further alternatives are machines based on computational intelligence, e.g. the artificial neural networks. All of these machines need an efficient communication system between the single nanomachines or neurons, and a powerful communication system to the outside world. In addition these nanosystems should make use of self-structuring and self-assembling so the designer can concentrate on higher levels in the design flow. This step if of high importance if we deal with high system complexities.

The present strategies for designing systems are dominated by the ideas of microelectronics: Miniaturization of classical circuits yielding higher performances. Only this way will not be sufficient for nanoelectronics. Due to the large complexity of the structures we have to develop new methods for designing systems. The design methods must be compatible with the technologies. It is still open whether fundamentally new concepts as such quantum computing will come to fruition, and result in different applications.

Today there are some speculations that nanoelectronics is able to produce intelligent robots that are superior to human beings. Therefore some people see the risk of being enslaved by these robots. However, we have no ideas for most characteristics of men so the question of being enslaved by robots is still open. The future research activities will show if machines will have a consciousness and if they can be an intelligent autonomous entity.

## 16.2 Uncertainties in Nanosystems

Probability is inherent to information. Additionally, it is also a characteristic feature of hardware, and in general for software, too, if it is complex and inscrutable, see Fig. 16.2. The implementation of logic functions into software could be made more secure and more reliable than in hardware since only mathematical context has to be transferred. However, reality shows a reverse behavior due to shortcomings and errors in computer programs.

Numerical information can always be transformed in an equivalent network composed of single switches. These switches can be implemented in several physical structures. Because of economical reasons certain functions are implemented in software. This concept is an essential strategy for the implementation of present systems.

Uncertainty plays a crucial role for the behavior of switches. If these switches are based on many-particle systems the switching behavior is subject to distribution functions of the particle energy, normally the Boltzmann or Fermi distribution. Since the distribution function reaches to infinity, thermal energy can cause errors in the switches of a network. Therefore all switches will not function in an absolutely safe manner. The same uncertainty is applies to a switch in a one-particle system. In this case, the Schrödinger equation describes the behavior of a particle. Once again the material waves extend to

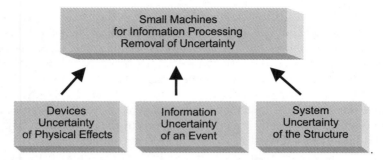

**Fig. 16.2.** Effects of uncertainties in an electronic system. The objective of these nanomachines is to remove uncertainties from our knowledge

infinity, and tunneling of particles can cause errors and diminish the reliability. The faulty behavior of the switches is usually compensated by the application of redundancy concepts, especially, these done in software.

The implementation of switches in physical structures causes reliability problems that intensively occur on the system level. The reliability is proportional to the probability, which indicates that a machine will work correctly during a certain time period, the so-called lifetime of a machine. Currently the demand for a long lifetime is achieved by careful selection of the physical components or switches with regard to their functionality and by special strategies in the development, for example by introducing redundant parts into the nanosystem. These problems increase with decreasing feature size of the devices because of the uncertain behavior of the switches. In addition, the number of failures grows exponentially with the number of switches or the complexity of the systems, as shown in Chap.15.

In future, engineers will have to learn how to build systems that are fault tolerant in spite of the presence of some defective devices. This is one reason why nanoelectronics needs fault-tolerant architectures. Artificial neural networks are one possible solution to overcome this problem. The implicit programming of such neural networks draws no conclusion of the network itself, i.e. the fault tolerance is not deterministic. Because of this other unexpected errors can occur. Therefore one challenge is to find an architecture that yields a perfect information-processing machine. The overall challenge is to build machines that should reliably remove uncertainties, whereas these machines themselves contain uncertain effects.

## 16.3 Uncertainties in the Development of Nanoelectronics

One main goal in the development of hardware is the realization of information-processing machines that are as small as possible, as mentioned above. The reasons for this strategy are high reliability, low costs, low power consumption, high performance, small volumes, and high processing speed. The presentation in Chap. 2 shows the limits of todays hardware implementations.

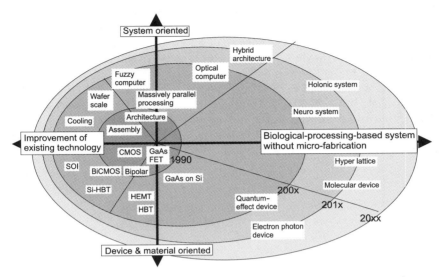

**Fig. 16.3.** Various ways for the development of electronic systems towards nanoelectronics

Today we have a vision of the concept for an information-processing system in nanoelectronics. Devices that are controlled by a few electrons or molecules correspond to single bits. Electrons are moving in molecular structures that are built by self-organization. This concept uses the same laws for the function of the devices as for the structuring of the nanocircuitries. At the moment this level of development is a vision and will not be achieved by technology soon.

This development can be done in many different ways, as Fig. 16.3 depicts. The key words in this scenario concern both the technology and the system architecture, and they have been explained in the preceding chapters. The distances between the ellipses concern the potential of development. For CMOS technology the potential will be exhausted soon, whereas nanotechnologies offer a wide field with really great progress.

## 16.4 Summary

The development of information systems has to be continued by several orders of magnitude as current forecasts predict. This goal can be met by introducing nanoelectronics. Nanoelectronics offer the potential of solving very complex problems by use of efficient parallel computing. In addition, nanoelectronics may offer new technologies that will substitute the very expensive silicon technology on a long-term basis.

The challenge of nanosystems concentrates on two fields: First, process engineers have to develop technologies that allow the manufacturing of microsystems with nanostructures at high yields, and also with a high reliability. Secondly, system engineers have to develop adequate architectures for ten billion or even more devices to provide an innovative information processing. Both problems are not solved today.

Without any doubt forecasts are questionable: The Nobel prize winning physicist Niels Bohr remarked: Forecasts are always difficult, especially those about the future. However, in the fast-growing field of electronics the remark of Noyce, the cofounder of Intel, is more relevant: It is much easier to forecast the future than it is to change the past [71].

# References

1. C. Mead, L. Conway: *Introduction to VLSI Systems* (Addison-Wesley, New York, 1980)
2. J. Meindl: The limits of semiconductor technology, Proc. IEEE **89**, pp.223–412 (2001)
3. K. Garbrecht, K. Stein: Perspectives and limitations of large scale systems, Siemens Forsch.-und Entw.Ber. pp. 312–318 (1976)
4. A. Grove: *Physics and Technology of Semiconductor Devices* (Wiley, New York, 1967)
5. Semiconductor Industries Association: "The International Technology Roadmap for Semiconductors 2001", http://public.itrs.net
6. M. Rudden, J. Wilson: *Elements of Solid State Physics* (Wiley, New York, 1993)
7. D. Widmann, H. Mader, H. Friedrich: *Technologie hochintegrierter Schaltungen* (Springer, Berlin Heidelberg New York, 1996)
8. W. Fahrner (Ed.): *Nanoprocessing and Nanoelectronics* (Springer, Berlin Heidelberg New York, 2003)
9. R. Compano, L. Molenkamp, D. Paul (Eds.): *Technology Roadmap for Nanoelectronics 2000* (European Commission IST Programme - Future and emerging Technologies, 2001), http://www.cordis.lu/ist/fethome.htm
10. Phantom: "Phantom: Nanotechnology in europe", http://www.phantomsnet.com/phantom/net/
11. R. Waser (Ed.): *Nanoelectronics and Information Technology, Advanced Materials and Novel Devices* (Wiley-VCH, Weinheim, Germany, 2003)
12. D. Nagel, M. Zaghloul: MEMS: Micro Technology, Mega Impact, IEEE Circuit and Devices pp. 14–25 (2001)
13. D. Paul: "Beyond CMOS: Si Heteroepitay and nanoelectronics for the superchip of the future", in *Mikroelektronik fuer die Informationstechnik* (VDE-Verlag, Berlin, 1998), no. 147, pp. 183–199
14. R. Feynman, R. Leighton, M. Sands: *The Feynman Lectures on Physics* (Addison-Wesley, Reading, 1964)
15. R. Colclaser, S. Diehl-Nagle: *Materials and Devices for Electrical Engineers and Physicists* (McGraw-Hill, New York, 1985)
16. L. Solymar, D. Walsh: *Lectures on the Electrical Properties of Materials* (Oxford University Press, Oxford, 1979)

17. S. Brandt, H. Dahmen: *The Picture Book of Quantum Mechanics* (Wiley, New York, 1985)
18. L. Brillouin: *Science and Information Theory* (Academic Press, New York, 1962)
19. C. Mead: *Analog VLSI and Neural Systems* (Addison-Wesley, New York, 1989)
20. C. Koch, I. Segev: *Methods in Neuronal Modeling* (MIT Press, Cambridge, 1989)
21. P. Fromherz, A. Offenhausser: A neuron-silicon junction: A retzius cell of the leech on an insulated-gate field-effect transistor, Science **252**, pp.1290–1293 (1991)
22. J. Delgado-Frias, W. Moore (Eds.): *VLSI for Artificial Intelligence and Neural Networks* (Kluwer Academic Publisher, Boston, 1991)
23. W. Young, B. Sheu: Unraveling the future of computing, IEEE Circuits and Devices **13**, pp.14–21 (1997)
24. M. Amos: *Theoretical and Experimental DNA Computation* (Springer, Berlin Heidelberg New York, 2003)
25. L. Adleman: Molecular computation of solutions to combinatorial problems, Science **266**, pp.1021–1024 (1994)
26. J. Gruska: *Quantum Computing* (McGraw-Hill, New York, 1999)
27. C. Williams, S. Clearwater: *Ultimate Zero and One, Computing at the Quantum Frontier* (Copernicus, 2000)
28. V. Tiwari: Instruction-level power analysis and optimization of software, J. VLSI Signal Proc. pp. 223–238 (1996)
29. C. Kozyrakis, D. Patterson: A new direction for computer architecture research, IEEE Comput. pp. 24–32 (1998)
30. H. Mattausch, K. Kishi, T. Gyohten: Area-efficient Multiport SRAMs for On-chip Data-Storage with High Random-Access Bandwidth and Large Storage Capacity, IEICE Trans. Electron. **E84-C**, pp.410–417 (2001)
31. S. Jung: *Capacitive CMOS Fingerprint Sensor with On-Chip Parallel Signal Processing* (VDE Verlag, Berlin, 2000)
32. T. Fountain: The use of nanoelectronic devices in highly parallel computing systems, IEEE Trans. VLSI Systems **6**, pp. 31–38 (1998)
33. J. Heath: A defect-tolerant computer architecture: Opportunities for nanotechnology, Science **280**, pp.1716–1721 (1998)
34. H. Schwefel, I. Wegener, K. Weinert (Eds.): *Advances in Computational Intelligence* (Springer, Berlin Heidelberg New York, 2002)
35. D. Fogel, T. Fukuda, L. Guan: Computational intelligence, Proc. IEEE **87**, 1415–1667 (1999)
36. J. Zurada, R. Marks, C. Robinson (Eds.): *Computational Intelligence, Imitating Life* (IEEE Press, New York, 1994)
37. A. Kramer: Array-based analog computation: Principles, advantages and limitations, Proc. of MicroNeuro96 pp. 68–79 (1996). IEEE
38. *Neural Assemblies* (Springer, Berlin Heidelberg New York, 1982)
39. *Self-Organizing Maps* (Springer, Berlin Heidelberg New York, 1997)
40. D. Slepian (Ed.): *The Development of Information Theory* (IEEE Press, New York, 1974)
41. D. Mange: Embryonics: A new methodology for designing field-programmable gate arrays with self-repair and self-replicating properties, IEEE Trans.VLSI Systems **6**, pp.387–399 (1998)

42. H. DeMan: Rethinking engineering research aand education for post-pc systems-on-a-chip, IEEE (2000)
43. IBM: "Picture serie", http://www.almaden.ibm.com/vis/stm/stm.html
44. L. Svensson: *Low Power Digital CMOS Design, Adiabatic Switching* (Kluwer Academic Publishers, Boston, 1995)
45. R. Keyes: *The Physics of VLSI Systems* (Addison-Weseley, New York, 1987)
46. R. Soref: Silicon-based optoelectronics, Proc. IEEE **81**, pp.1687–1706 (1993)
47. J. Singh: *Semiconductor Devices, An Introduction* (McGraw-Hill, New York, 1994)
48. K. Hoffmann: *Systemintegration* (Oldenbourg, Munich, 2003)
49. N. Weste, K. Eshraghian: *Principles of CMOS VLSI Design* (Addison-Wesley, New York, 1992)
50. V. Beiu, J. Quintana, M. Avedillo: VLSI Implementations of Threshold Logic - A Survey, IEEE Solid State Circuits (to be published)
51. R. Marrian: Nanometer-scale science and technology, Proc. IEEE **85** (1997)
52. G. Wirth, U. Hilleringmann, J. Horstmann, K. Goser: Mesoscopic Transport Phenomena in Ultrashort Channel MOSFETs, Solid-State Electron. **43**, pp.1245–1250 (1999)
53. A. Seabaugh, P. Mazumder: Quantum devices and their applications, Proc. IEEE **87** (1999)
54. R. Merkle: (1991), "Computational nanotechnology", ftp://parcftp.xerox.com/pub/merkle/merklesHomePage.html
55. J. Ellenbogen: (1999), "Architectures for molecular electronic computers", http://www.mitre.org/technology/nanotech
56. K. Maezawa, T. Mizutani: A New Resonant Tunneling Logic Gate Employing Monostable-Bistable Transition, Jpn. J. Appl. Phys. Part 2 **32**, pp.42–44 (1993)
57. P. Mazumder, S. Kulkarni, M. Bhattacharya, J. Sun, G. Haddad: Digital Circuit Applications of Resonant Tunneling Devices, Proc. IEEE **86**, pp.664–686 (1998)
58. K. Goser, C. Pacha: "System and circuit aspects of nanoelectronics (invited paper)", in *Proceedings of the 24th European Solid-State Circuits Conference ESSCIRC* (1998), pp. 18–29
59. C. Pacha, U. Auer, C. Burwick, P. Glösekötter, W. Prost, F.J. Tegude, K. Goser: Threshold logic circuit design of parallel adders using resonant tunneling devices, IEEE Trans. VLSI Syst. **8**, pp.558–572 (2000)
60. P. Glösekötter, W. Prost, C. Pacha, H.v.H. S.O. Kim, T. Reimann, F.J. Tegude, K. Goser: "Pseudo dynamic gate design based on the resonant-tunneling-bipolar-transistor", in *32nd European Solid-State Device Research Conference, ESSDERC*, Florence, Italy (2002)
61. P. Hadley: Single electronics: One electron, one bit?, FED J. **4**, pp.20–27 (1994)
62. R. Klunder: (2002), "Circuit design with metallic single-electron tunnel junctions", Ph.D. thesis, TU Delft
63. H. Ahmed: Single electron electronics: Challenge for nanofabrication, J. Vac. Sci. Technol. **B,15**, pp.2101–2108 (1997)
64. D. Averin, K. Likharev: *Single Charge Tunneling - Possible Applications of the Single Charge Tunneling* (Plenum Press, New York, 1992), Chap. 9, pp. 311–332
65. N. Asahi, M. Akazawa, Y. Amenmiya: Single electron logic devices, IEEE Trans. Electron Devices **44**, pp.1109–1116 (1997)
66. C. Wasshuber, H. Kosina, S. Selberherr: A comparative study of single-electron memories, IEEE Trans. Electron Devices **45**, pp.2365–2371 (1998)

67. R. Keyes: Scaling, small numbers and randomness in semiconductor, IEEE Circuits and Devices pp. 29–31 (1994)
68. F. Reynolds: Thermally accelerated aging of semiconductor components, Proc. IEEE **62**, pp.212–222 (1974)
69. L.B. Kish: End of Mooret's Law: Thermal (Noise) Death of Integration in Micro and Nano Electronics, Phys. Let. A **305**, pp.144–149 (2002)
70. R. Keyes: Physical uncertainty and information, IEEE Trans. Comput. pp. 1017–1025 (1977)
71. G. Heilmeier: "Technology scenarios for the information age", in *Interface 89 Symposium Paris* (1989)

# Index

$\Psi$ function, 188

active transistor area, 28
adder, 145
address bus, 104
adiabatic switching, 136
AFM, 21
aggressive scaling, 250
AND gate, 87
anisotropic selective etching, 31
area time product, 97
Arrhenius Plot, 260
ASIC, 6
associative matrix, 91, 118
associative memory, 232
atomic force microscope, 21
autonomy, 108

bacteriorhodopsin, 183
band diagram, 13, 153
basic logic circuits, 198
baud, 57
BCS theory, 227
best matching unit, 120
biochemical computers, 77
bioelectronics, 169
biomolecular computer, 82
bipolar transistor, 1
Boltzmann distribution, 49
Boltzmann factor, 135
Boltzmann law, 66
Bose'Einstein distribution, 49
bremsstrahlung, 48
brilliant system, 10

Broglie wave, 211

C60 molecules, 176
cellular automata, 117
cellular neural network, 117
center-of-gravity method, 110
characteristic time, 4
charge carriers, 40
charge coupled devices, 144
chip area, 246
classical set theory, 109
CMOS technology, 143
coherence, 85
coherent superposition, 85
compiler, 127
complexity, 267
complexity of integrated nanoelectronic systems, 7
computational intelligence, 107
conditioning, 73
conducting molecules, 197
conduction band, 14
connectionistic systems, 113
connective systems, 90
content-addressable memory, 232
convolution systems, 97
Cooper pairs, 227
Coulomb blockade, 40, 209, 219
Coulomb gap, 212
Coulomb's law, 209
crossover operation, 112
cryogenic conduction, 137
cryotron, 228
CUPS, 115

data memory, 89
de Broglie wave, 193
de Broglie wavelength, 39
Debye length, 4, 17, 152, 257
defuzzyfication, 110
degree of parallelism, 95
depletion-layer width, 4
device-failure mechanisms, 260
dielectric breakdown, 28
diffusion constant, 50
diffusion process, 252
diffusion voltage, 65
DMP, 174
DNA computer, 77
DNA strand, 79
DNA polymerase, 79
donor-acceptor-bridge, 174
doping concentration, 255
doping density, 4
double minimun potential, 174
double helix, 78
Drexler and Merkle, 172
dynamic circuit design technique, 143
dynamic logic gates, 199
dynamic power consumption, 96

E(x)-diagram, 13
economics, 7
EEPROM, 190
effective mass, 15
Einstein, 171
electric breakdown, 17
electric design rules, 127
electron beam writing, 19
electron pump, 216
electronic cochlea, 72
electronic eye, 138
electronic retina, 72
energy bands, 13
energy supply, 248
entanglement, 86
entropy, 9, 54
EPR experiment, 84
equal-probability distribution, 252
error-tolerant implementations, 101
Esaki diode, 191
evolutionary hardware concepts, 129
explicit complexity, 8

failure rates, 265
fault tolerance, 92, 130
fault-tolerant architectures, 262, 270
Fermi level, 15, 157
Fermi'Dirac distribution, 49
fine structure, 18
fine structure constant, 42
fingerprint recognition, 100
floating bar, 33
flux quantum, 225
flux-quanta, 42
Fredkin gate, 126, 147
fullerenes, 176
functional integration, 3
fuzzy logic, 107
fuzzy systems, 108

gas sensor, 33
gate oxide thickness, 17
gate-channel-capacitance, 28
gel electrophoresis, 80
genetic algorithms, 112
genetic information, 78
geometric design rules, 127
graph algorithms, 97

HBT, 205
heat conduction, 248
heat dissipation, 248
Hebbian rule, 62, 72, 114
Heisenberg's uncertainty principle, 44
heterojunction bipolar transistor, 205
HFET MOBILE, 205
HFET-RTD MOBILE, 204
Hilbert space, 85
Hopfield Network, 166
hot electrons, 27
hybrid technique, 35

ideal switch, 134
implicit complexity, 8
information, 43
information content, 8, 52, 56
information cuboid, 57
information processing machines, 267
information quantity, 57
inherent bistability, 200
inhomogeneous doping profile, 257
insulation materials, 28
integration level, 2

intelligent dust, 37
intelligent RAM, 98
IRAM, 98

Josephson elements, 87
Josephson tunneling device, 229

Klitzing structure, 158

Langmuir-Blodgett, 170
LC logic, 234
leakage currents, 18
ligases, 80
light-emitting diode, 35
limits, 3
linear programming, 97
linear threshold gate, 202
linguistic rules, 107
linguistic unsharpness, 108
lithography process, 19
load balancing, 90
local data processing, 72
local processing, 117
localized reconfigurability, 101
logic entropy, 125
look-up table, 102
LTG, 202

M. Forshaw, 137, 141
machine code, 127
macroscopic electrical effects, 39
magnetic flux quantum, 234
majority gate, 153
massive parallel information processing, 82
matrix switches, 104
Maxwell equations, 39
MBE, 19
mean effective velocity, 14
mean free path, 5
mean time between failure, 261
Meißner-Ochsenfeld, 225
memory applications, 198
MEMS, 31
MESFET, 15
mesoscopic device, 197
mesoscopic elements, 152
mesoscopic switching element, 187
microelectronic and mechanical system, 31

microlaser, 35
microminiaturization, 19
microstructure effects, 27
minimum feature size, 17
MIPS, 10, 57
MOBILE, 199
mobility, 14
molecular beam epitaxy, 19, 156
molecular diode, 178
molecular electronics, 34
monostable-bistable transition, 204
Moore Plot, 2
MTBF, 261
multiport memories, 99
multiprocessor system, 89
multivalued data storage, 219
multivibrator, 69
mutation, 112

nanoelectronic interface, 32, 127
nanoelectronics, 3, 152
nanometric switching devices, 137
nanoprinting, 21
NDR, 187
negative differential resistance, 187
negative differential resistance, 190
negentropy, 8, 56, 125
Neumann principle, 89
neuron, 73
neuronal network, 61
nonlinear I-V characteristics, 187
NP complete problem, 8
nucleases, 80
numerical information, 267
Nyquist theorem, 258

OEIC, 35
one-dimensional potential well, 45
optoelectronic integrated circuit, 35

parallel processing, 10, 77
parameter spread, 252
particle momentum, 44
PDN, 205
periodic potential, 13
photonics, 34
photons, 40
physical limits, 245
PIP, 100

280    Index

pipeline architectures, 92
Planck's constant, 5
pn-junction, 14
Poisson equation, 257
potential barrier, 142, 188, 190
potential energy, 188
power dissipation, 249
power-delay characteristic, 24
power-delay diagram, 136, 220, 250
power-dissipation density, 246
Poynting vector, 40
program memory, 89
propagated instruction processor, 100
pseudo dynamic, 199
pull-down network, 205

q-switch, 159
QCA, 160, 234
QED, 6
quanta, 41
quantization, 209
quantum electronic device, 262
quantum barriers, 16
quantum cellular arrays, 84
quantum cellular automata, 234
quantum computer, 134, 166, 234, 263
quantum computing, 83, 245
quantum confinement, 211
quantum cryptography, 84
quantum dot, 155, 211
quantum effects, 6
quantum electronic devices, 6
quantum Hall effect, 242
quantum layers, 154
quantum noise, 47
quantum well, 16, 152, 193
quantum wire, 155
quantum-dot architectures, 187
quantum-dot structures, 152
quantum-effect device, 187
quantum-mechanical limit, 251, 265
quantum-mechanical tunneling effect, 28
Qubit, 85, 234, 236

random unsharpness, 108
rapid single flux quantum device, 237
real switch, 134
reconfigurable circuitries, 101

reconfigurable optical network, 139
redox-cycling-based electrochemical sensor, 173
redundancy, 262
reference vector, 120
relativistic limits, 264
relay, 141
reliability, 259
removal of uncertainties, 267
replacement of technologies, 246
resonant tunneling bipolar transistor, 196
resonant tunneling diode, 192
RISC, 98
RSFQD, 237
RTBT, 196
RTBT MOBILE, 204
RTBT multiplexer, 206
RTBT threshold gate, 205
RTD, 19, 187, 192
RTD-based threshold gate, 203
RTD-HFET, 200

salicyl-aniline, 175
scaling, 23
scanning tunneling microscope, 21
Schottky contact, 15
Schrödinger equation, 43
Schrödinger's wavefunction, 188
selective reproduction, 82
self organizing principles, 75
self-assembling, 181
self-organization, 120, 181
self-organizing maps, 120
self-stabilizing state, 200
semantic information, 267
Semiconductor Industries Association, 6
semiconductor laser, 35
sensor systems, 18
SET, 152, 187, 196, 209
SET circuit design, 216
SET technology, 214
SFQ-logic, 231
SFQD, 236
shift register, 216
short-channel MOS transistor, 18
SIA, 6
silent servant, 10
silicon on insulator, 17

silicon technology, 246
SIMD architecture, 95
single electron transistor, 40
single flux quantum device, 236
single probe methods, 21
single-electron transistors, 187, 196, 209, 251
single-port memory, 99
singletons, 110
six-port memory, 99
soft failures, 8
softcomputing, 107
SOI, 17, 214
solid-state nanoelectronics, 169
space-charge region, 191, 254
spatial entropy, 125
spatial fluctuation, 257
spintronics, 159
SPM, 21
SQUID, 233
stimulated transition, 48
Stirling's equation, 55
STM, 21
structural dimension, 4
subthreshold domain, 69
subthreshold voltage, 71
superconducting computer, 236
superconducting quantum interferometer device, 233
superconductivity, 225
supercurrent, 42
switching energy, 29
systolic array, 91
systolic array algorithms, 97
systolic arrays, 90

tactile molecular processor, 171
TD, 187, 190
TE, 152, 187
technological limits, 246
technology of RTD, 196
Teramac concept, 102
testing, 7
the electrical standards, 241
thermal energy, 26
thermal fluctiations, 135
thermal noise, 258, 265
three-dimensional integration, 3, 18
three-dimensional signal processing, 138
three-port memory, 99

three-terminal resonant tunneling device, 196
three-terminal tunneling devices, 187
threshold gate, 63, 115
threshold value, 64
thyristor, 142
total power consumption, 95
transistor
  bipolar, 142
  MOS, 141
  single-electron, 152
  split-gate, 157
transitional energy, 66
transmissible power, 258
transmission line, 27
traveling salesman problem, 8, 80
trend curves, 2
tunnel effect, 13, 188
tunneling, 265
tunneling devices, 187
tunneling diode, 15, 142, 187, 190
tunneling element, 187
tunneling elements, 152
tunneling probability, 188
tunneling resistance, 213
Turing machine, 57, 79

ubiquitous networking, 267
ubiquitous computing, 267
uncertainties in nanosystems, 269
uncertainties in the development of nanoelectronics, 270
uncertainty, 267
uncertainty principle, 213
unitary transformation, 263

valence band, 14
veto-input, 73
via hole, 18
viologene, 175
von-Klitzing resistance, 214

wavefunction, 188
wavelength, 5
wiring, 8

X-radiation lithography, 20
XOR problem, 115

yield, 255